Atmospheric Physics

This view of the Earth – extending from the Mediterranean Sea area to the Antarctic cap – was photographed by Apollo 17 crewmen during their journey toward the Moon. (NA

Atmospheric Physics

J. V. Iribarne and H.-R. Cho

McLennan Physical Laboratories, University of Toronto, Canada

D. Reidel Publishing Company

Dordrecht : Holland / Boston : U.S.A. / London : England

Library of Congress Cataloging in Publication Data

CIP

Iribarne, Julio Víctor.
Atmospheric physics.

Bibliography: p.
Includes index.
1.   Atmosphere. I. Cho, Han-Ru, 1945–   joint
author. II. Title.
QC861.2.I74   551.5   79-23098
ISBN-13: 978-94-009-8954-2   e-ISBN-13:978-94-009-8952-8
DOI: 10.1007/978-94-009-8952-8

Published by D. Reidel Publishing Company,
P.O. Box 17, 3300 AA Dordrecht, Holland

Sold and distributed in the U.S.A. and Canada
by Kluwer Boston, Inc., Lincoln Building,
160 Old Derby Street, Hingham, MA 02043, U.S.A.

In all other countries, sold and distributed
by Kluwer Academic Publishers Group,
P.O. Box 322, 3300 AH Dordrecht, Holland

D. Reidel Publishing Company is a member of the Kluwer Group

Table of Contents

Preface

The extraordinary growth and development of atmospheric sciences during the last decades, and the concern for certain applied problems, such as those related to the environment, have prompted the introduction of college and university courses in this field. There is consequently a need for good textbooks.

A few appropriate books have appeared in the last few years, aimed at a variety of levels and having different orientations. Most of them are of rather limited scope; in particular, a number of them are restricted to the field of dynamics and its meteorological applications. There is still a need for an elementary, yet comprehensive, survey of the terrestrial atmosphere. This short volume attempts to fill that need.

This book is intended as a textbook that can be used for a university course at a second or third year level. It requires only elementary mathematics and such knowledge of physics as should be acquired in most first-year general physics courses. It may serve in two ways. A general review of the field is provided for students who work or plan to work in other fields (such as geophysics, geography, environmental sciences, space research), but are interested in acquiring general information; at the same time, it may serve as a general and elementary introduction for students who will later specialize in some area of atmospheric science. The book is quite comprehensive, as a perusal of the table of contents will indicate; for instance, chapters on chemistry and electricity – usually absent in this type of book – have been included. Indeed, it tries to introduce most of the basic concepts and facts about the atmosphere that do not require any previous specialization. The variety and organization of the included topics are such that a lecturer can always choose to exclude specific subjects described, without losing continuity. Lists of questions and problems have been added to each chapter; these, even if elementary in nature, should help the reader to acquire some insight into the various subjects.

In summarizing from different sources, we owed much to such excellent elementary texts as those by Dobson and by Goody and Walker, as also to monographs written at a higher level. A short bibliography with some brief comments has been included; we hope that it may provide useful orientation toward further reading.

We are indebted to Profs. R. List and C. Hines and Dr. R. E. Munn for kindly reading and commenting on parts of the manuscript. For facilitating the use of photographic material, we would like to express our thanks to Prof. R. List (Toronto), Prof. R. Montalbetti (Saskatoon), Prof. G. Soulage (Clermont-Ferrand), Dr. J. Joss and Mr. H. Binz (FKH, Switzerland), Prof. T. Fujita (Chicago), Drs. Ch. and N. Knight (NCAR), scientists of the Atmospheric Environment Service of Canada (in particular Dr. R. Schemenauer and Messrs. Ch. Taggart and A. A. Aldunate), EROS Data Center, and NASA. Thanks are also due to Mrs. C. Banic for her help in the preparation of the collection of problems, and to Miss J. Cooper for her efficient secretarial assistance.

Acknowledgements

The authors thank all concerned for permission to reproduce the following figures.

Ch.I, 1, 5, 11, Ch.II, 3, Ch.III, 3: G. M. B. Dobson, *Exploring the Atmosphere* (Oxford University Press).

Ch.I, 9, 10: Prof. R. Montalbetti, University of Saskatchewan.

Ch.I, 12: James A. Van Allen, 'Radiation Belts Around the Earth', *Scientific American* (March 1959).

Ch.III, 5: C. O. Hines et al. (eds.), *Physics of the Earth's Upper Atmosphere* (Prentice-Hall).

Ch.III, 8: M. J. McEwan and L. P. Phillips, *Chemistry of the Atmosphere* (Edward Arnold).

Ch.III, 12, 13: J. C. Johnson, *Physical Meteorology* (MIT Press).

Ch.III, 14: R. E. Newell, 'The Circulation of the Upper Atmosphere', *Scientific American* (March 1964).

Ch.V, 1, 2, 3, 4, 5, 19, 21: Atmospheric Environment Service, Toronto; also Ch.VII, 12, 32, 34, received from U.S. NOAA Weather Satellite.

Ch.V, 6, Ch.VII, 13: NASA, Houston.

Ch.V, 8: J. E. McDonald, *Advances in Geophys.*, **5**, 223 (1958).

Ch.V, 16, 18: Prof. G. Soulage, Université de Clermont.

Ch.V, 20, 22, 24: Prof. R. List, University of Toronto.

Ch.V, 23: Photographic Department of the National Center for Atmospheric Research, by courtesy of Dr. Ch. Knight.

Ch.VI, 12, 13: Blitzmessstation der FKH, Monte San Salvatore, Switzerland, by courtesy of the Forschungskommision des SEV und VSE für Hochspannungsfragen (FKH), Zürich.

Ch.VI, 15: M. A. Uman, *Lightning* (McGraw-Hill).

Ch.VII, 18, 22, 26, 27: E. Palmén and C. W. Newton, *Atmospheric Circulation Systems* (Academic Press).

Ch.VII, 22a: E. Palmén and L. A. Vuorela, *Quart. J. Roy. Meteorol. Soc.*, **87**, 131–138 (1963).

Ch.VII, 22b: L. A. Vuorela and I. Tuominen, *Pure Appl. Geophys.*, **57**, 167–180 (1964).

Ch.VII, 23: Y. Mintz, *Bull. Amer. Meteorol. Soc.*, **57**, 208–214 (1954).

Ch.VII, 24: European Space Agency, Paris.

Ch.VII, 26: F. Defant and H. Taba, *Tellus*, **9**, 259–274 (1957).

Ch.VII, 27: D. L. Bradbury and E. Palmén, *Bull Amer. Meteorol. Soc.*, **34**, 56–62 (1953).

Ch.VII, 33, 35: Canadian Meteorological Centre, Montreal.

Table of Constants

Fundamental Physical Constants

Gas constant	$R = 8.314\,\text{J/mol.K}$
Avogadro's number	$N_A = 6.02 \times 10^{23}\,\text{mol}^{-1}$
Boltzmann's constant	$k = 1.38 \times 10^{-23}\,\text{J/K}$
Planck's constant	$h = 6.63 \times 10^{-34}\,\text{J.s}$
Velocity of light in vacuum	$c = 3.00 \times 10^{8}\,\text{m/s}$
Elementary charge	$e = 1.60 \times 10^{-19}\,\text{C}$
Gravitational constant	$G = 6.67 \times 10^{-11}\,\text{Nm}^2/\text{kg}^2$

Earth

Standard acceleration of gravity	$g_0 = 9.81\,\text{m/s}^2$
Solar constant	$S = 2.0\,\text{cal/cm}^2.\text{min} \simeq 1400\,\text{W/m}^2$
Earth's mean radius	$R = 6.37 \times 10^{6}\,\text{m}$
Earth's surface area	$5.1 \times 10^{14}\,\text{m}^2$
Mean Earth–Sun distance	$1.49 \times 10^{8}\,\text{km}$
Mean Earth–Moon distance	$3.80 \times 10^{5}\,\text{km}$
Angular velocity of rotation	$\Omega = 7.29 \times 10^{-5}\,\text{s}^{-1}$

Atmosphere

Standard atmospheric pressure	$1\,\text{atm} = 1.01325 \times 10^{5}\,\text{Pa}$
Total mass of atmospheric air	$5.3 \times 10^{18}\,\text{kg}$
equivalent to	$4.1 \times 10^{18}\,\text{m}^3$ at STP ($M = 28.964\,\text{g/mol}$)
Average molecular weight of atmospheric air, up to 100 km altitude	$M = 28.964\,(\text{g/mol})$
Dry adiabatic temperature lapse rate	$\beta_d = 9.76\,\text{K/km}$

Atomic and Molecular Weights*

H	: 1.01	C	: 12.01
He	: 4.00	N	: 14.00
Ne	: 20.18	O	: 16.00
Ar	: 39.95	S	: 32.06

* These constants enter the formulas as conversion factors with units of mass per mole. E.g. $M_{H_2O} = 18.0\,\text{g/mol} = 0.018\,\text{kg/mol}$.

Air   : see above
H_2O : 18.02
NaCl : 58.44
CO_2 : 44.01

SO_2 : 64.06
H_2S : 34.08
NH_3 : 17.02

Thermodynamic

Heat capacity of air, at constant pressure	$C_p = 29.1$ J/mol.K $= 1005$ J/kg.K
Heat capacity of air, at constant volume	$C_v = 20.8$ J/mol.K $= 718$ J/kg.K
Heat capacity of water	$C_w = 76$ J/mol.K $= 4218$ J/kg.K
Latent heat of melting (water) $(0°C)$	$L_f = 6.01 \times 10^3$ J/mol
Latent heat of vaporization (water) $(0°C)$	$L_v = 4.50 \times 10^4$ J/mol
Latent heat of sublimation (water) $(0°C)$	$L_s = 5.10 \times 10^4$ J/mol

Saturation water vapour pressures:

        At $0°C$:   6.11 mb

    At other temperatures: See Figure II, 1

Others

Density of water	10^3 kg/m^3
Density of dry air at $0°C$ and 1 atm (STP)	1.29 kg/m^3
Surface tension of water: $0°C$	$\sigma = 0.0756$ N/m
$20°C$	$\sigma = 0.0727$ N/m
Viscosity of air $(0°C)$	$\eta = 1.71 \times 10^{-5}$ Ns/m^2
Permittivity of free space	$\epsilon_0 = 8.854 \times 10^{-12}$ F/m
Permeability of free space	$\mu_0 = 4\pi \times 10^{-7}$ H/m
Stefan–Boltzmann constant	$\sigma = 56.7$ nW/m^2 K^4
Mass of the electron	9.11×10^{-31} kg

Units

The International System (SI) is used in general. This implies MKS mechanical units (based on the fundamental units metre, kilogram and second for length, mass and time, respectively), Kelvin (K) for absolute temperature, Ampere (A) for electric current, and mole (mol) as a chemical unit of mass.

Other units used are summarized below.

Temperature:
Degrees Celsius ($^{\circ}$C), defined by $t = T - 273.15$, where t is the temperature in $^{\circ}$C and T is the absolute temperature in K.

Pressure:
Atmosphere (atm). 1 atm $= 1.013\,25 \times 10^{5}\,$Pa, where Pa $=$ Pascal $= $ N/m^2 (N $=$ Newton) is the SI unit.
Millibar (mb). 1 mb $= 100\,$Pa.

Energy:
Calorie (cal). 1 cal $= 4.184\,$J, where J (Joule) is the SI unit.
Electron-volt (eV). 1 eV $= 1.6 \times 10^{-19}\,$J.
(If the energy is referred to one mole, 1 eV corresponds to 96.3 kJ/mol).

The SI system uses prefixes to indicate a multiplying factor. Some names, symbols and corresponding factors are given in the following table:

tera (T)	: $\times 10^{12}$	milli (m) : $\times 10^{-3}$
giga (G)	: $\times 10^{9}$	micro (μ): $\times 10^{-6}$
mega (M)	: $\times 10^{6}$	nano (n) : $\times 10^{-9}$
kilo (k)	: $\times 10^{3}$	pico (p) : $\times 10^{-12}$

I. General Description of the Atmosphere

1. Regions and Extension of the Atmosphere

The atmosphere is a gaseous envelope surrounding the Earth, held by gravity, having its maximum density just above the solid surface and becoming gradually thinner with distance from the ground, until it finally becomes indistinguishable from the interplanetary gas.

There is, therefore, no defined upper limit or 'top' of the atmosphere. As we go away from the surface of the Earth, different regions can be defined, with widely different properties, being the seats of a great variety of physical and chemical phenomena. If we want to understand the atmosphere, our first concern will be to introduce some sort of classification that will help to consider separately all these phenomena. Let us consider the pictorial representation of Figure 1.

First of all, let us observe that, apart from the first kilometre, the scale used is a logarithmic one, so that the upper regions are in fact much thicker, in comparison with the lower ones, than suggested by the picture.

There is a scale of pressures at the right side. The pressure, at each level, is given by the weight of all the air above it, per unit area of surface, and this weight is given by

$$\int_z^\infty g \, \rho \, dz$$

where ρ = density, g = acceleration of gravity, z = height (mass times acceleration of gravity, integrated above the level z). The value of g varies only slowly with height. Therefore, the pressure can be taken as roughly proportional to $\int_z^\infty \rho \, dz$, i.e. to the total mass above that level z. We can see, by comparing the pressure scale with the height scale at the left, that:

> 90% of the mass is contained within the first ~ 20 km (top at 100 mb level)
> 99.9% of the mass is contained within the first ~ 50 km (top at 1 mb level)

At 100 km, the pressure has dropped to about 10^{-3} mb, i.e. only a fraction of the order of 10^{-6} (one millionth) of the atmospheric mass will be above that level. And only a fraction 10^{-13} above 1000 km. These distances are to be compared with the Earth radius, ~ 6370 km. It is clear that from the point of view of its mass, the atmospheric envelope, although of diffuse limits, is a very thin sheath around the planet.

2. Homosphere and Heterosphere. Scale Height

The 'thickness' of the atmosphere can be characterized by a parameter called the *scale height*, which we shall now define. We can consider the atmosphere essentially as a fluid

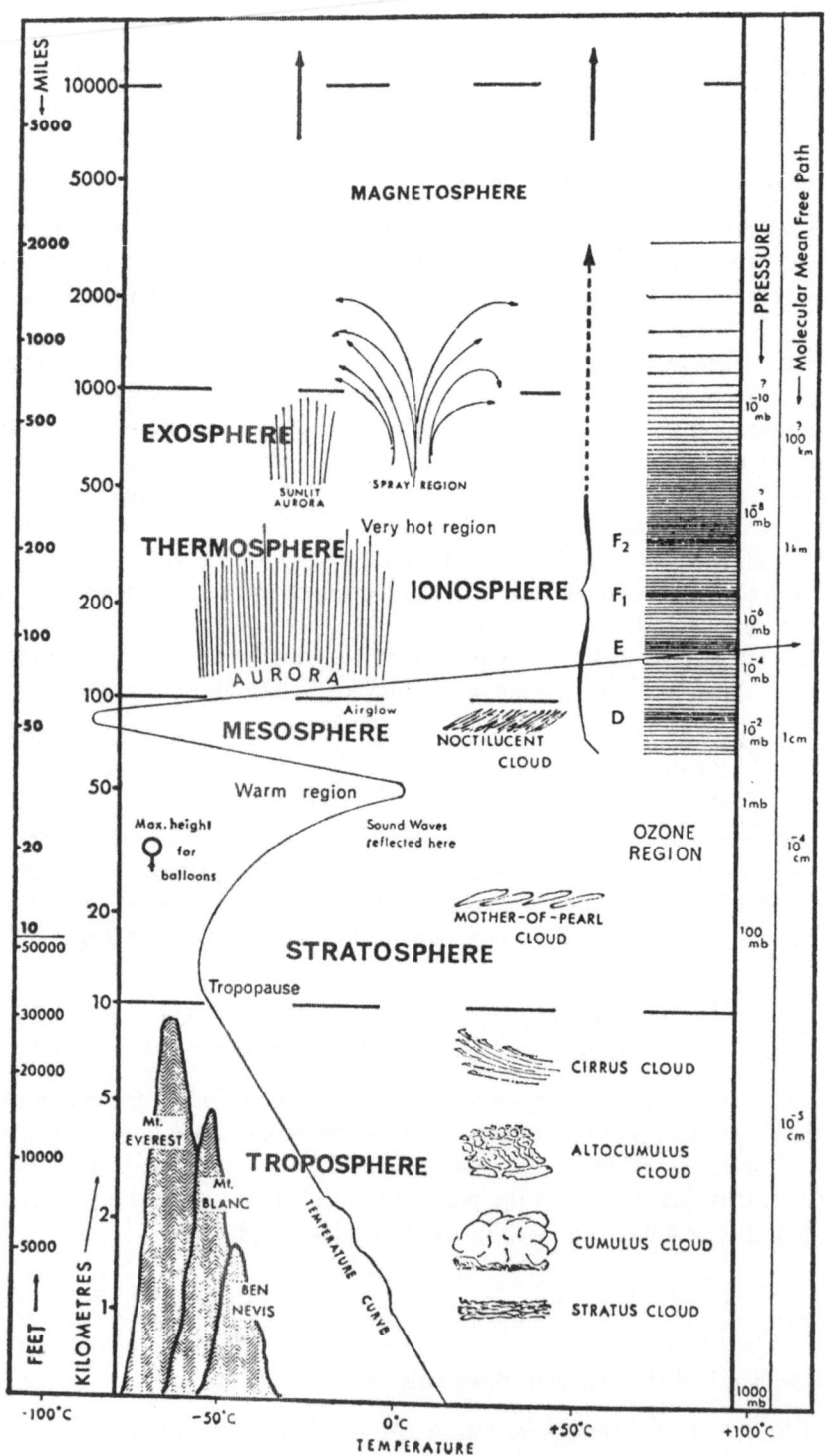

Fig. I-1. *The atmosphere.* Pictorial representation showing the main features of the terrestrial atmosphere. The explanation of this figure is developed in Ch. I.

---

in hydrostatic equilibrium. This means that for every infinitesimal layer of unit cross section we shall have the following relation (cf. Figure 2):

$$\text{upward force due to pressure gradient} = \text{weight}$$

$$-\mathrm{d}p = g \cdot \rho \cdot \mathrm{d}z \qquad (1)$$

This relation (1) (where p = pressure) is called the *hydrostatic equation*.

We observe now that air, at the temperatures and pressures of the atmosphere, behaves like a mixture of ideal gases, within at most a few percent of error. For lower and lower pressures, the behaviour approaches more and more that of an ideal gas. For each component we shall then have

$$p_i V = n_i RT \qquad (2)$$

where p_i = partial pressure of component i; V = volume; n_i = number of moles of component i; R = universal gas constant; T = absolute temperature.

And for the mixture

$$p = \Sigma p_i \qquad \text{(Dalton's law)}$$

$$pV = nRT \qquad (3)$$

where p = total pressure and n = total number of moles. An average molecular weight is defined as $M = m/n = (\Sigma n_i M_i)/n$ where m = mass and M_i = molecular weight of component i.

Equation (3) can also be written

$$p = \frac{RT}{M}\rho \qquad (4)$$

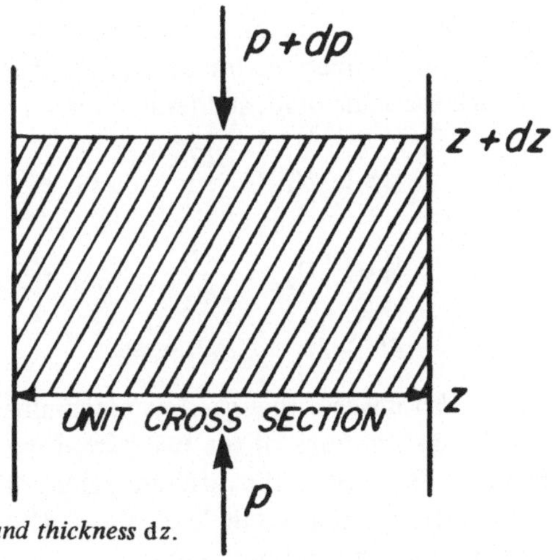

Fig. I-2. *Layer of air column of unit cross section and thickness* dz.

Eliminating ρ between (1) and (4), we can express the hydrostatic equation as

$$d \ln p = -\frac{gM}{RT} dz \tag{5}$$

which can be integrated to

$$p = p_0 \exp\left(-\int_0^z \frac{gM}{RT} dz\right) = p_0 \exp\left(-\int_0^z \frac{dz}{H}\right) \tag{6}$$

where we have defined the parameter $H = RT/gM$, which we call the (local) scale height of the atmosphere. As we shall see, M can be considered as a constant up to 100 km; g depends on z, but it varies only about 3% for every 100 km. H therefore varies roughly proportionally to T up to 100 km. For $T = 273$ K, we obtain ($g = 9.8$ m/s^2; $M = 28.96$)

$$H = 8 \text{ km} \tag{7}$$

So, if the temperature was uniform, (6) could be written

$$p = p_0 \cdot e^{-z/H} \tag{8}$$

and H would indicate the height at which the pressure has decreased by a factor $e^{-1} = 0.37$, i.e. the height within which about $\frac{2}{3}$ of the atmosphere mass would be contained. Actually, as T varies with height, so does H; but up to 100 km, only within the range of 5 to 9 km.

So far, we have assumed that the atmospheric air is a mixture of gases of constant composition, perfectly mixed, so that M is a constant in Equation (6). But this is not the situation throughout the whole atmosphere. Let us consider the problem more closely. If we have a mixture of gases in the gravitational field and wait for the equilibrium distribution, the Statistical Mechanics predicts that there will be a separate distribution for each kind of molecules; i.e. each gas will obey its own equation:

$$p_i = p_{0_i} \exp\left(-\int_0^z \frac{dz}{H_i}\right) \tag{9}$$

according to its own value of M. This leads to a predominance of the heavier molecules in the lower levels, and of the lighter molecules in the higher levels. To understand this clearly, let us imagine, for simplicity, that g and T are constant, so that H varies only through the value of M, and let us consider a mixture of only two gases.

$$\text{gas 1, } H_1 = \frac{RT}{gM_1} \quad , \quad p_1 = p_{0_2} e^{-z/H_1}$$

$$\text{gas 2, } H_2 = \frac{RT}{gM_2} \quad , \quad p_2 = p_{0_2} e^{-z/H_2}$$

$$M_1 > M_2 \qquad H_1 < H_2$$

Let us also imagine that gas 1 is predominant at the surface. Then the two distributions, i.e. the distributions of the two partial pressures p_1 and p_2, will be such as is shown in Figure 3. Because of the different values of H in the exponent in Equation (9), the two curves intersect at a certain level, above which the lighter gas 2 becomes predominant.

The reason we did not consider this in the previous argument is that this type of equi-

Fig. I-3. *Height distribution of gases in diffusive equilibrium.* p = partial pressure; z = height over reference level; p_{0_1}, p_{0_2} = partial pressure of gas 1, 2 at $z = 0$; H_1, H_2 = scale height of gas 1, 2. H_1 and H_2 have been assumed constant, so that the two curves are exponential.

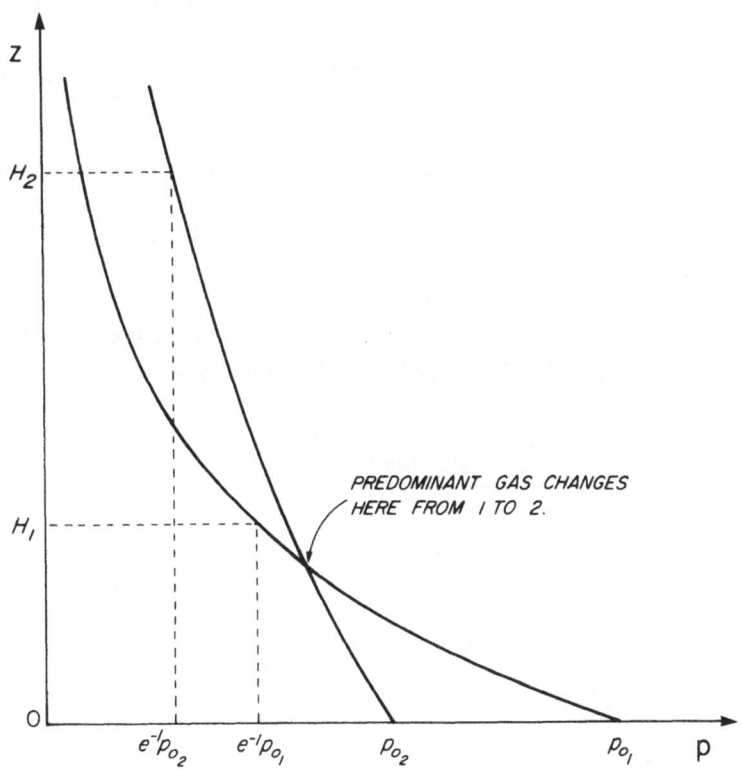

librium takes very long to be reached. Under 100 km mixing mechanisms (presumably turbulence and convective overturning) are active enough to keep the atmospheric air well mixed. The hydrostatic distribution, on the other hand, becomes established very quickly and at every instant we can assume that it is basically existent. Above 100 km the situation changes: there is essentially no more mixing. Therefore, under 100 km the air composition, except for trace gases, is constant and (6) is applicable to the mixture; above 100 km, the equilibrium should better be described by the set of equations (9), and the composition starts varying accordingly. This is still far from being the whole picture, however, as other factors are also modifying this composition at these levels, mainly the photochemical action of solar radiation. But we shall leave consideration of this aspect for later on. Here we shall only retain this feature: that the layer below ~ 100 km is of constant chemical composition (except for water vapour and certain trace gases) and is therefore called the *homosphere*, while the part above ~ 100 km is of varying composition and for that reason is called the *heterosphere*.

Thus we have come to a first classification of regions of the atmosphere, based on chemical composition. In this respect we should also mention a trace gas of particular importance, ozone, which forms photochemically in the stratosphere, where it has a maximum concentration in the region of 20 to 30 km. This ozone layer, which absorbs

5

the short wave range of solar radiation and is responsible for a maximum of temperature at about 50 km, is sometimes known as the *ozonosphere* (cf. Ch.III, §6).

There is one important gas which is very variable in the lower layers of the atmosphere: water vapour. We shall study it later on. Here we only mention that from its condensation we obtain in the troposphere (lower layers; see below) the different kinds of clouds, whose typical altitudes are represented for reference in Figure 1. All the usual clouds appear in the troposphere, but occasionally clouds can be observed in two higher levels. These are the *mother-of-pearl clouds*, so called because of their iridescent aspect, which appear in a narrow range of height, around 27 km, and possibly consist of frozen droplets rather than crystals; and the *noctilucent* clouds, which consist of ice crystals probably nucleated over meteor dust, and appear at 80–100 km, visible after sunset, while still illuminated by the Sun.

It may be interesting to compare the scale height of the Earth (typical value) with the scale heights in other planets, where both the composition of the atmosphere and the gravitational acceleration are different. Such comparison is given in Table 1.

TABLE 1[†]

Scale heights of planetary atmospheres

	Gas	Mean molecular mass (amu)[*]	Gravitational acceleration (cm sec^{-2})	Average surface temperature (K)	Scale height (km)
Venus	CO_2	44	888	700	14.9
Earth	N_2, O_2	29	981	288	8.4
Mars	CO_2	44	373	210	10.6
Jupiter	H_2	2[**]	2620	160[***]	25.3

[†] Reproduced from Goody and Walker (see Bibliography).
[*] Atomic mass units.
[**] There may be enough He (4 amu) on Jupiter to increase the mean molecular mass significantly and hence to decrease the scale height.
[***] The temperature near the top of Jupiter's clouds.

Other classifications of the atmospheric regions than that based on chemical composition are also possible, based on temperature distribution, on electron density, etc. The classification based on temperature distribution is particularly important, and we shall consider it now.

3. Temperature Distribution: Troposphere, Stratosphere, Mesosphere, Thermosphere

Figure 1 shows a curve representing schematically a typical distribution of temperatures in height. Starting from the ground and up to a certain height, the temperature normally decreases at a rate of 5 to 7 degrees per km. This is variable with time and place, and even occasionally there occur shallow layers within which the temperature increases with height: the so-called *inversions*. We shall later come back on this question of the temperature distribution and its relation with vertical stability (Ch.IV). The region under consideration is called the *troposphere* and is the seat of the weather phenomena that affect us at the ground. It is also, for obvious reasons, the best known region, and it contains about $\frac{4}{5}$ of the total air mass. Its upper limit is defined by a sudden change in the temperature trend, often appearing as a discontinuity in the curve: the temperature stops decreasing more or less suddenly and remains constant or starts increasing slightly. This limit is called the *tropopause* and its height, also depending on time and place (being larger at the Equator than at the Poles) can vary between 7 and 17 km. The temperature at the tropopause in middle latitudes is − 50 to − 55°C.

The next region shows a gradual increase of temperature, reaching a maximum of around 0°C at 50 km. The region is called the *stratosphere* and its upper limit (at the temperature maximum), the *stratopause*. Then the temperature drops again through the region called the *mesosphere*, to a minimum of the order of − 100°C defining the *mesopause*, at about 85 km. From there on the temperature increases steadily and the region is called the *thermosphere*. The temperature, of whose increase only the beginning is shown in Figure 1, reaches high values and then remains constant; at 500 km it may reach values between 400 and 2000°C, depending on the time of day, degree of solar activity and latitude; the diurnal variation is of 500–800°C, with minimum near sunrise and maximum at about 2 p.m. Figure 4 indicates the type of temperature distribution, on a natural height scale. Notice that these high temperatures in the thermosphere do not imply that

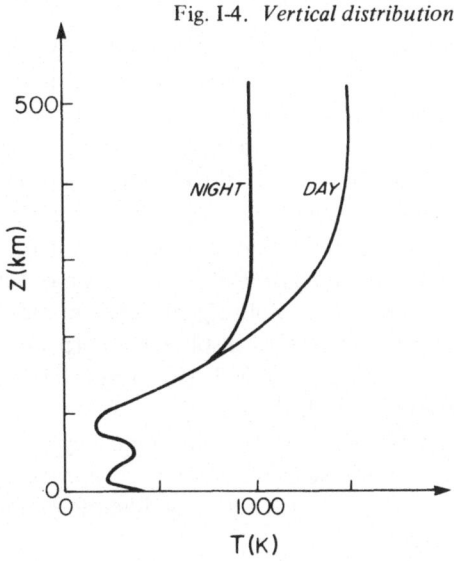

Fig. I-4. *Vertical distribution of temperature* (typical curves; see text).

an object crossing this region, such as a satellite, may be much affected, because the pressure is too low (10^{-8} mb at 500 km) to permit any appreciable heat transfer; the object will be actually in very high vacuum.* The meaning of the temperature at these

---

* The temperature of a satellite will be determined by radiation exchange.

pressures is best grasped by considering that it gives a measure of the average kinetic energy of the molecules.

The physical reasons for this peculiar distribution of temperature in the atmosphere are related to the absorption of radiation, to be treated in Ch.III. Reactions of ionization and dissociation occur in the upper levels, producing the high temperatures of the thermosphere. The maximum at the stratopause is associated with the presence of the ozone. The ground is again normally at a maximum, due to the absorption at the surface of a large fraction of the remainder of solar radiation that reaches that level.

In order to complete the description of Figure 1, we must now refer to some problems of physics of the upper atmosphere. We shall do so with some more detail than necessary at this stage, because we shall not develop this subject area any further in the rest of the book, except as concerns the chapter on radiation. The physics of the lower layers of the atmosphere, on the other hand, will be treated more fully in the following chapters.

4. Ionosphere

At high levels, the shorter wavelengths of solar radiation produce a considerable amount of ionized atoms and molecules and the corresponding free electrons. Thus a vertical distribution of electron density exists, with characteristic layers or regions of high density, associated with, and depending on, the photochemical processes; the typical distributions are therefore different for night and day conditions. These regions, when considered from this point of view, are called the *ionosphere*.

The ionosphere can be conveniently studied from the ground by radiowaves in the range of 1–20 MHz (300–15 m of wavelength). The instrument used is called an *ionosonde*, and consists basically of a pulse transmitter and a receiver. The transmitter emits short pulses of a given frequency v vertically upwards, as indicated schematically in Figure 5. When this electromagnetic radiation passes through the ionosphere, the free electrons oscillate in response to the electromagnetic field of the wave. It may be shown that, for a high concentration of electrons, reflection takes place. Part of the reflected signal is received back at the ionosonde, which registers it, acting as a receiver. The whole process takes a time of the order of a millisecond. Both signals, the emitted pulse and the echo, are displayed on a cathode ray tube as vertical displacements (peaks) over a horizontal time basis, so that the time lag can be determined. This time interval indicates the double distance (upwards and downwards) travelled by the pulse. When automatic sweeping over frequency is provided, the horizontal axis gives the frequency and it is arranged that the height of the signal on the screen for each frequency is proportional to the delay (i.e. indicates height of reflection); a record of such a display, like the example shown in Figure 6, is called an *ionogram*.

The height at which the beam is reflected depends both on the frequency and on the electron number concentration N; from the theory of wave propagation, it may be derived that the condition for reflection is

$$N = \frac{4\pi^2 \epsilon_0 m v^2}{e^2} = \text{const} \times v^2 \tag{10}$$

where ϵ_0 = free-space permittivity ($\epsilon_0 = 8.854 \times 10^{-12}$ F/m in the SI system), m = mass

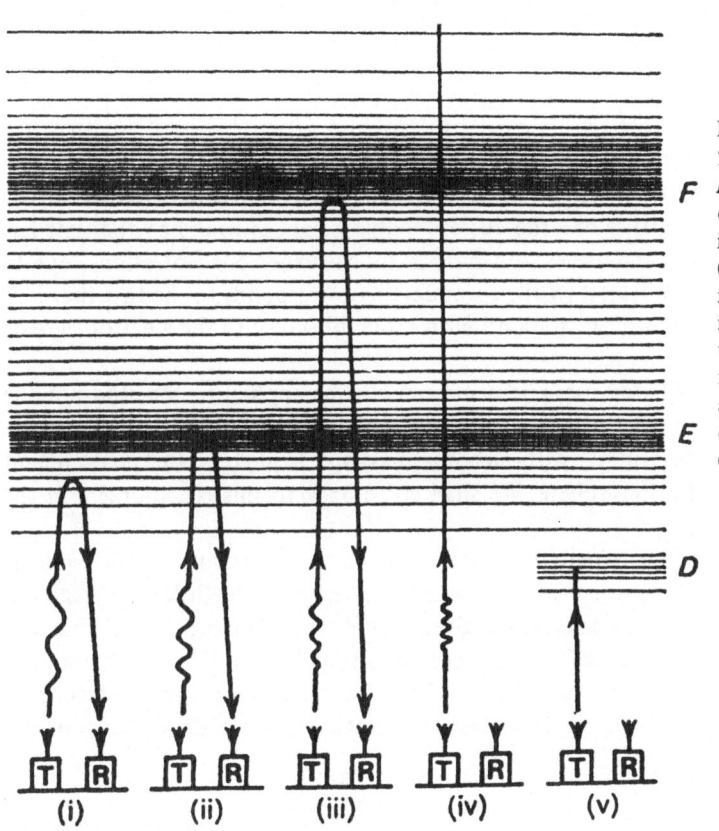

Fig. I-5. *Ionosonde.*
T indicates the transmitter and
R the receiver. The intensity
of ionization in the ionosphere
is indicated by shading. (*i*), (*ii*),
(*iii*) and (*iv*) represent the
reflection of pulses of increasing frequency. (*v*) illustrates
the fact that the *D* region
must be investigated separately, because it is more difficult to detect, due to the lower
electron density.

Fig. I-6. *Ionogram.* A record showing the height at which a pulse of electromagnetic wave is reflected
as a function of pulse frequency. The ionogram is produced by a device consisting of an automatically-
sweeping pulse transmitter and receiver, called 'ionosonde'. The maxima of the signal correspond to the
peaks of various regions of the ionosphere. The small decrease of the signal after each maximum and
the doubling of the signals are due to properties of wave propagation not considered in the elementary
explanation presented in the text. Frequencies are indicated in megacycles per second, heights in
kilometres.

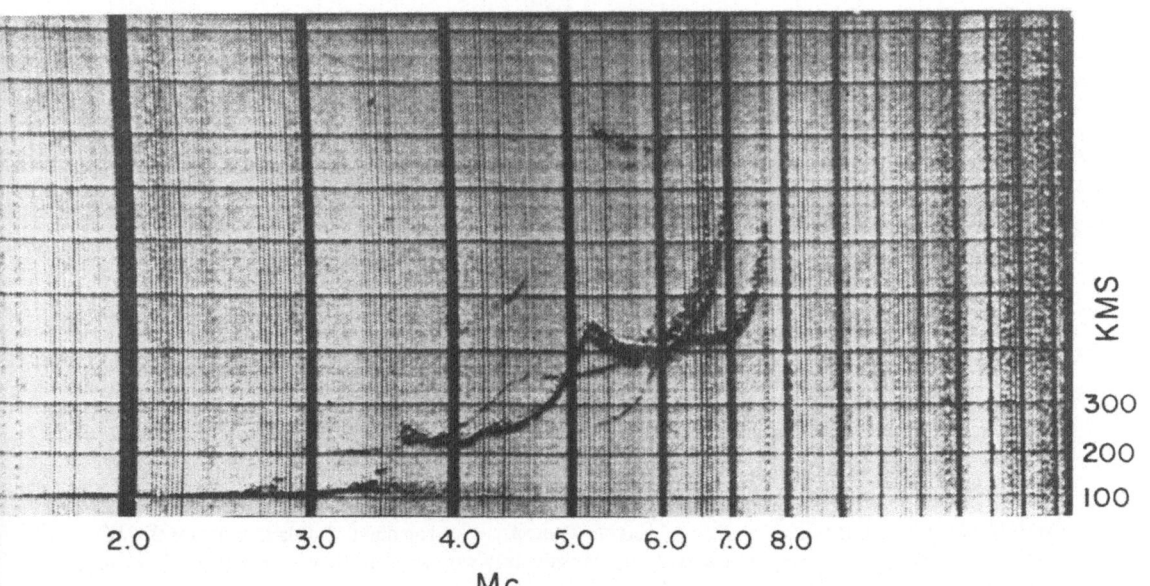

of the electron, e = charge of the electron. If N is given in cm^{-3}, the proportionality constant has the value 1.24×10^{-8} s^2.cm^{-3}. For each frequency the concentration N can be derived from Equation (10), while the corresponding height is found by the time lag. The ions would also reflect the waves, but only at much higher concentrations, due to their much larger mass, as indicated by the formula; this corresponds to the fact that ions are much less mobile, and therefore respond with more difficulty to the electromagnetic field of the waves.

Other less direct ground-based measurements can also be applied to the study of the ionosphere. More important is the use, in recent times, of rockets and satellites to measure electron densities and to determine, with airborne mass spectrometers, the predominant ionic species present in the ionosphere. We shall come back to this last problem in the chapter on Radiation (§5).

The experimental results obtained with these techniques are summarized in Table 2. Four regions have been distinguished in the ionosphere, designated with D, E, F_1 and F_2; their height ranges and electron densities are given in the table. The region D is observed during daytime only, and the regions F_1 and F_2 become a single region F (or F_1 disappears) during the night. This is illustrated in Figure 7, where the electron densities are given in a logarithmic scale against height. This figure also illustrates the direct dependence of the

TABLE 2
Ionospheric regions

Region	Heights (km)	Electron density (cm^{-3}) (typical order of magnitude)
D	< 90	10^3–10^4
E	90–140	10^5
F_1 F_2	> 140	Maximum of 10^6 in the region of 250–500 km

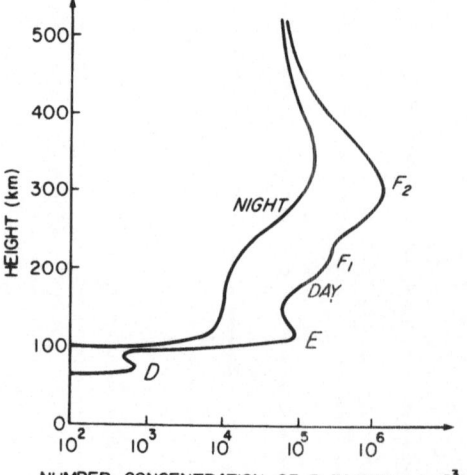

Fig. I-7. *Electron density as a function of height.* Typical curves for night and day, showing the different regions D, E, F_1, and F_2. The curves vary considerably with solar activity, sun-spot number and time of the year.

10

ionosphere on the ionizing radiation coming from the Sun. As well as this diurnal variation, an annual variation can be discerned.

The D region, occurring at lower altitudes, has a higher concentration of molecules, and this leads to some characteristics different from those of higher layers. The electron oscillations produced by electromagnetic waves cause frequent collisions with molecules. As a consequence, the energy imparted to the electrons by the wave can be rather easily dissipated, i.e. transferred to the molecules; instead of reflecting the waves, the region tends to absorb them in large extent (their energy being transformed into heat). This fact makes the study of this layer difficult, and is the cause of poor daytime reception of commercial long-wave radio transmission from distant broadcasting stations, as these waves are reflected in higher layers and must cross the D region twice. The effect virtually disappears at night, with the D layer.

A second consequence of the relatively high frequency of electron-molecule collisions in the D region is that many electrons become attached to the molecules, forming negative ions, at least by night. This process is very dependent on pressure because, as will be explained later (see Ch.III, §5), it requires not only the capturing molecule and the electron, but also the presence of a third body (another molecule or atom) during the collision, to carry away the excess energy produced in the reaction. As a result, the D region contains negative ions, which the other ionospheric regions do not, as well as positive ions and free electrons.

At the level of the E and F regions, the mean free path of free electrons is much larger, and collisions with molecules are infrequent. Therefore these layers consist of positive ions and free electrons; negative ions can be entirely neglected.

At all heights below 6000 km, even at the maximum in the F region, the number density of electrons is less than that of neutral particles. This is illustrated in Figure 8, where 'Total particles' includes neutral molecules, ions and electrons.

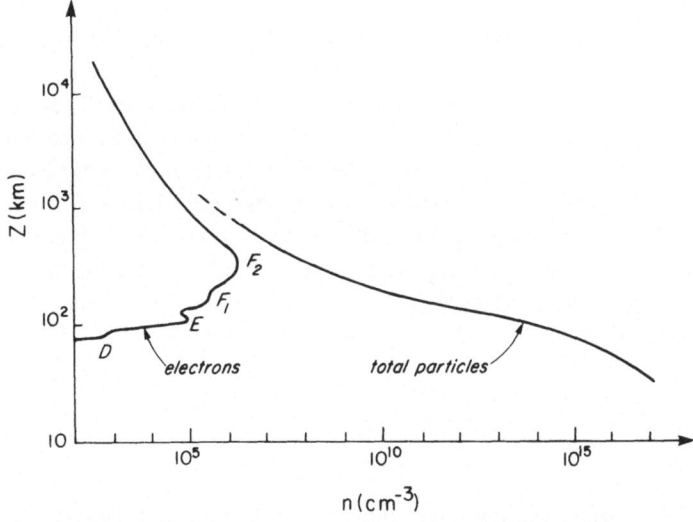

Fig. I-8. *Electron and total particle densities as functions of height.* The figure shows that electron density is always smaller than that of neutral particles (which make up most of the 'total particles'). For comparison, the molecular (= total particle) concentration at the ground is about 3×10^{19} cm^{-3}.

11

We mentioned before the periodical variations of the ionosphere. This referred to the normal or 'quiet' conditions (or with a 'quiet Sun'). However, at certain times the conditions become disturbed; we speak then of a 'disturbed' or 'active' Sun. These conditions are associated with Sun spots and 'solar flares'.

The solar flares are extremely hot masses of gas that leave the chromosphere* of the sun in an arching path and return to the surface. These flares are the source of intense ultraviolet radiation and of particles, mainly electrons and protons. The electromagnetic radiation reaches the terrestrial atmosphere first, in about 8 minutes, causing *sudden ionospheric disturbances* (SID), which are short-lived abnormalities occurring on the day-lit hemisphere and which result from a marked increase in the electron density of the lower ionosphere. The particles from the flare are very energetic; travelling at high speeds, they may arrive as soon as 20 minutes after the electromagnetic radiation and they penetrate down to the D region. The enhanced ionization produced by these particles is responsible for the black-out of radio communications, due to the associated enhanced absorption of the waves crossing that region.

We leave further discussion of the processes of ionization in various conditions and the ion composition in the ionosphere for Ch.III.

5. Exosphere

At the right of Figure 1, a scale indicates the approximate values of the *mean free paths*, λ, of molecules. This parameter is defined as the average distance travelled by a molecule between two collisions. It can be derived from the kinetic theory of gases that

$$\lambda = \frac{1}{\sqrt{2}} \frac{1}{\sigma n} = \frac{kT}{\sqrt{2}\sigma p} \tag{11}$$

where σ = collision cross section of the molecules (of the order $40\,\text{Å}^2 = 0.4\,\text{nm}^2$), n = number concentration, $k = 1.38 \times 10^{-23}\,J/K$ = Boltzmann constant, T = absolute temperature, p = pressure. Mainly because p becomes small very rapidly, and also because of the increase in T, λ increases from 10^{-5} cm at the surface to tens of kilometers at 500 km. This value is so large that collisions between molecules virtually cease at these heights, and the molecules or atoms perform parabolic (ballistic) trajectories in the gravitational field.** This region, above 500 km, that marks the gradual transition from the terrestrial atmosphere to the interplanetary gas, is called the *exosphere*. Its base is also called the 'spray region', suggesting a comparison between the molecular trajectories with those of water droplets coming out of the top of a vertical water jet.

---

* The region surrounding the central disc of the Sun, or photosphere.
** The base of the exosphere is actually defined as the level for which λ becomes equal to the scale height H. Above this level, for a particle travelling upwards, λ increases (mainly through the decrease in p) so rapidly that further collisions become less likely.

6. Aurora

Aurora is a luminous phenomenon observed as a glow coming from the upper atmosphere during the night in winter, in high latitudes: the *aurora borealis* in the northern hemisphere, and the corresponding *aurora australis* in the southern hemisphere. Figures 9 and 10 give examples of aurora. Most auroras are observed in a belt around the geomagnetic pole, between $15°$ and $30°$ from it, with a maximum frequency at about $22.5°$. They appear in a wide variety of patterns. An international classification based on their morphology exists and has been illustrated with the photographs of the International Auroral Atlas*; the classification is largely based on the work of Carl Störmer. The main divisions are band-like forms, diffuse forms, rays and other forms; further identification is made by using descriptive names, such as draperies, corona, quiet arc, pulsating arc, homogeneous bands, pulsating homogeneous patches, etc.

A typical display lasts for a period of about half an hour, with maximum activity during a few minutes. It develops rapidly and sometimes shows apparent motions of high velocities.

The theory of aurora is not yet fully understood. It seems clear that the phenomenon is associated with disturbed conditions in the upper atmosphere. The glow results from the bombardment and subsequent ionization of the molecules by particles coming from above. The ionized air molecules, on recapturing electrons and readjusting to their normal energy states, emit light of given frequencies, a type of process that will be explained in Ch.III. This bombardment is related to a variety of solar disturbances producing charged particles that arrive to the Earth, penetrate into its atmosphere and interact in a complicated manner with the Earth's magnetic field. The main wavelengths found in the auroral light are 557.7 nm, corresponding to a green line of the spectrum of atomic oxygen and 636.3 and 630.0 nm, corresponding to a red doublet line of oxygen. As might be expected, many lines and bands of nitrogen have also been identified, but except for a red band of nitrogen at 650 nm they are much weaker than the two oxygen lines. The resulting luminosity is usually bright green.

The height at which auroras develop is mainly from 80 to 150 km, although they may also occur at higher altitudes.

At high latitudes, in summer months, aurora is visible in the sunlit portion of the upper atmosphere after ground sunset or before ground sunrise. It is usually white-violet or blue-violet and appears at great heights (cf. 'sunlit aurora' in Figure 1).

The *airglow*, mentioned in Figure 1, is a weak luminosity always present. It has been studied mostly at night, but it is also present, with somewhat different characteristics, during day and twilight. Although not completely understood, it seems that most of the energy emitted during night glow derives from solar radiation which has been absorbed during the day and which becomes gradually available through subsequent chemical reactions as chemiluminescence. The green line and the red doublet of oxygen mentioned above are present, apparently having their origin in emissions from different height regions. Other frequencies have been identified as originated by the radical hydroxyl OH and by sodium atoms.

---

* Published for the International Union of Geodesy and Geophysics, Edinburgh, at the University Press.

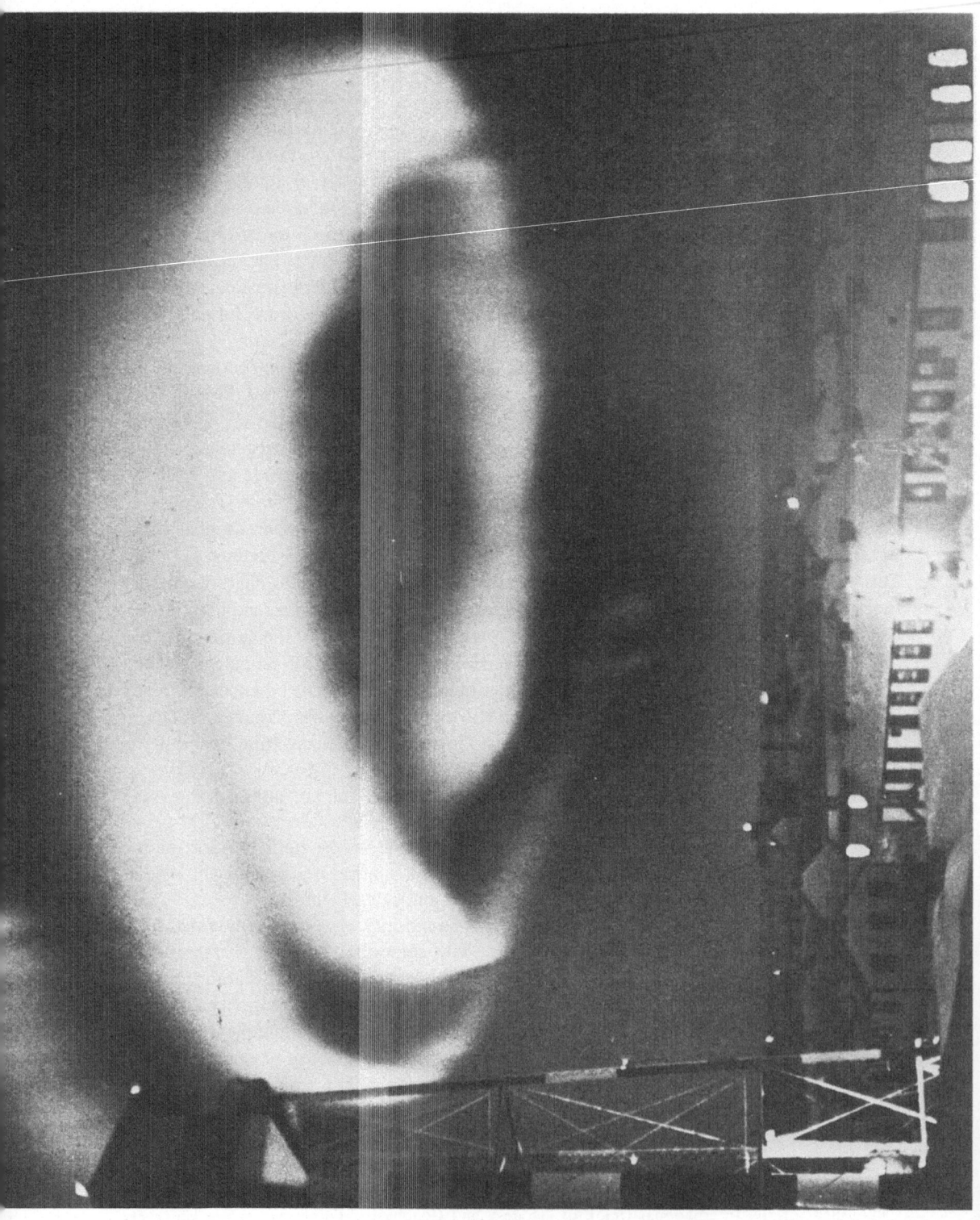

Fig. I-10. *Aurora borealis*. Band, with some ray structure. Photograph taken at Saskatoon, Canada.

7. Magnetosphere

There still remains a region included in the pictorial representation of Figure 1. It refers to the magnetic properties of the Earth and their interaction with the continous stream of corpuscular solar radiation that fills the interplanetary space, i.e. the so-called *solar wind*. Figure 11(a) shows the magnetic field around the Earth as it would be if the Earth were isolated in space. The charged particles (both positive particles and electrons) of the solar wind are deflected by the Earth's magnetic field with a force perpendicular to both the magnetic field and the trajectory of the particles:

$$\mathbf{F} = q\mathbf{v} \times \mathbf{B} \qquad \text{(SI system)} \tag{12}$$

where $\mathbf{F}$ = force acting on the charged particles; q = charge; $\mathbf{v}$ = velocity; $\mathbf{B}$ = magnetic induction. The effect of $\mathbf{B}$ is to deflect the particles back, away from the Earth. There is a reaction on the magnetic field, which becomes then modified as shown in Figure 11(b).

Fig. I-11. *Magnetosphere.* (a) The figure shows diagrammatically the magnetic field of the Earth as it would be if the Earth were isolated in space. The Earth is supposed to be viewed from a point on the magnetic equator. Distances are measured in terms of the Earth's radius as the unit; (b) the figures show the magnetic field of the Earth as in Figure 11(a), but now allowing for the effect of the solar wind. The double line indicates the magnetopause. The approximate positions of the Van Allen belts are shaded. For simplicity, the solar wind is shown as blowing at right angles to the Earth's magnetic axis, but this angle will vary.

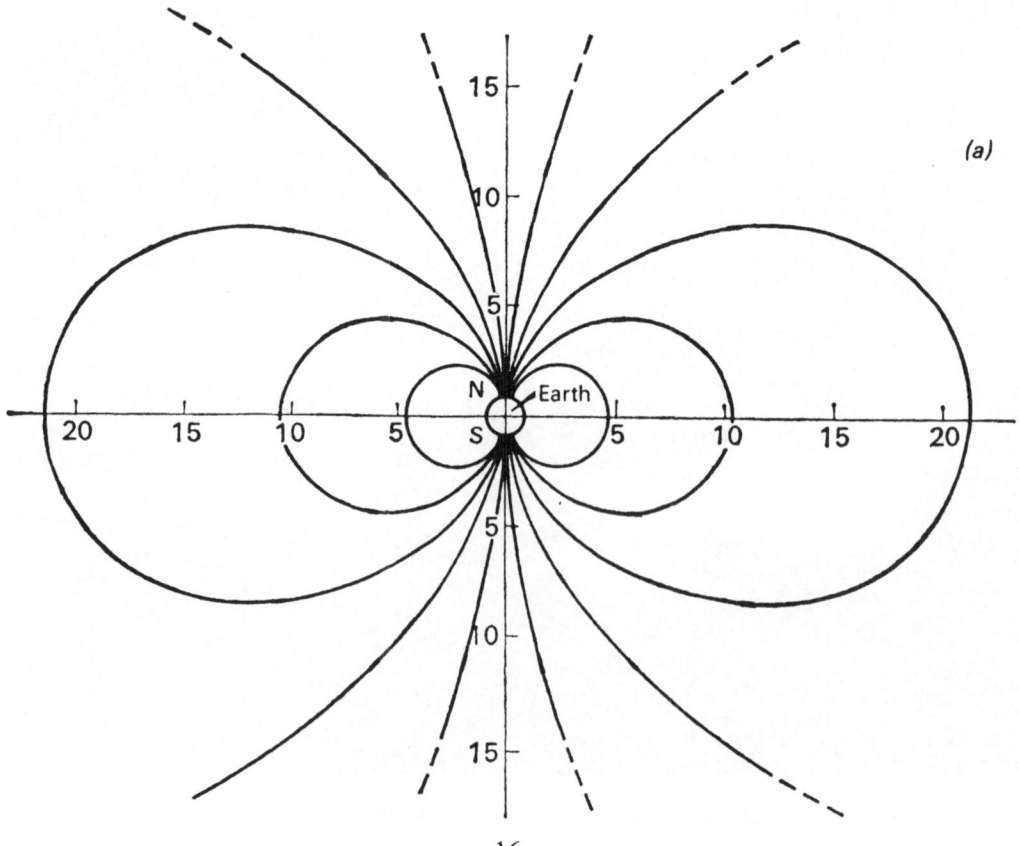

At a distance of 10–15 Earth radii on the daylight side (1 Earth radius = 6370 km) the field drops to 0 or negligible value. This limit is called the *magnetopause* and all the region within it, the *magnetosphere*. On the night side (away from the Sun), the magnetosphere extends to far greater distances; this is known as the Earth's magnetic tail.

Although the particles are in general rejected by the field, collisions may disturb the paths and the particles may become trapped in the magnetic field of the Earth. When this happens, the particle will start spiralling around a line of force, i.e. will describe a corkscrew-shaped trajectory having a magnetic line of force at its axis (see Figure 12). The turns of the spiral will be more open at the equator and will become tighter as the particle reaches stronger magnetic field toward the poles. At a certain point ('mirror point') the spiral becomes flat and then the particle winds back along a similar path towards the other hemisphere. Close to the other magnetic pole, the same process occurs, so that the particle spirals back and forth from one hemisphere to the other, in time periods of the order of the second. During this time, the axis of the trajectory shifts slightly, so that the particle drifts slowly around the Earth at the same time as it spirals back and forth. Electrons drift from west to east; protons, in the opposite direction. At each end of the path, the particles descend to regions of higher density, where the probability of collisions with atoms or molecules of the air becomes higher. Thus, after a period of days or weeks, a particle may undergo a collision, which reduces its energy; the particle is thereby removed from the magnetosphere and falls toward the lower atmosphere.

These trapped particles concentrate in two main regions around the Earth, the so-called *Van Allen belts* (from the name of their discoverer), which are indicated by a shading in Figure 11(b).

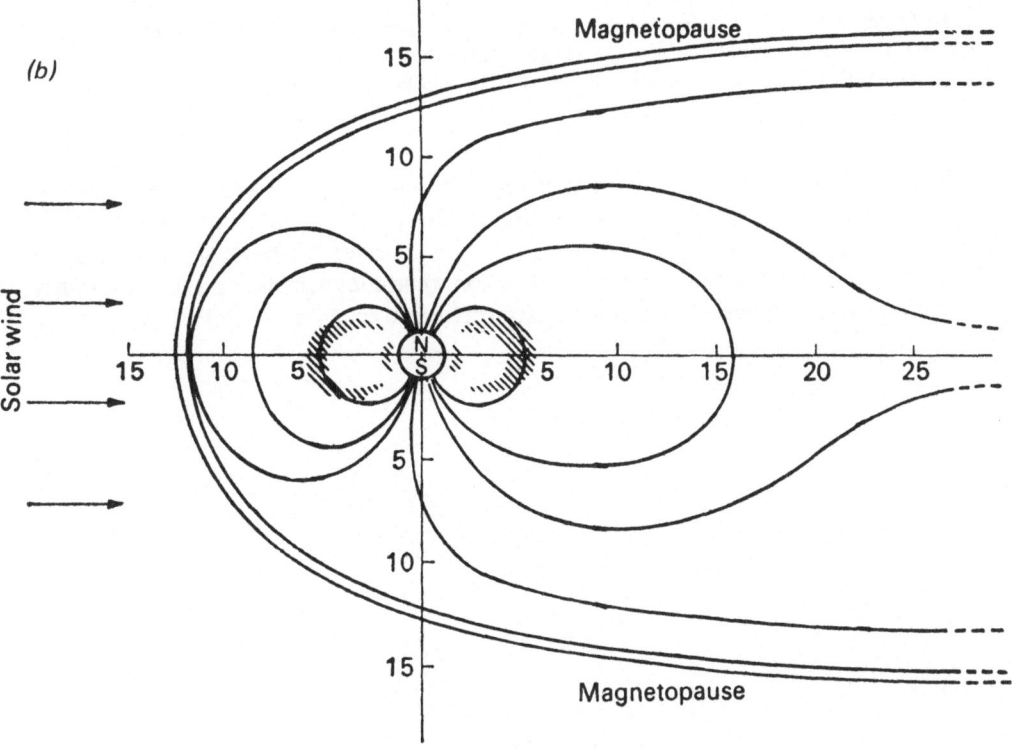

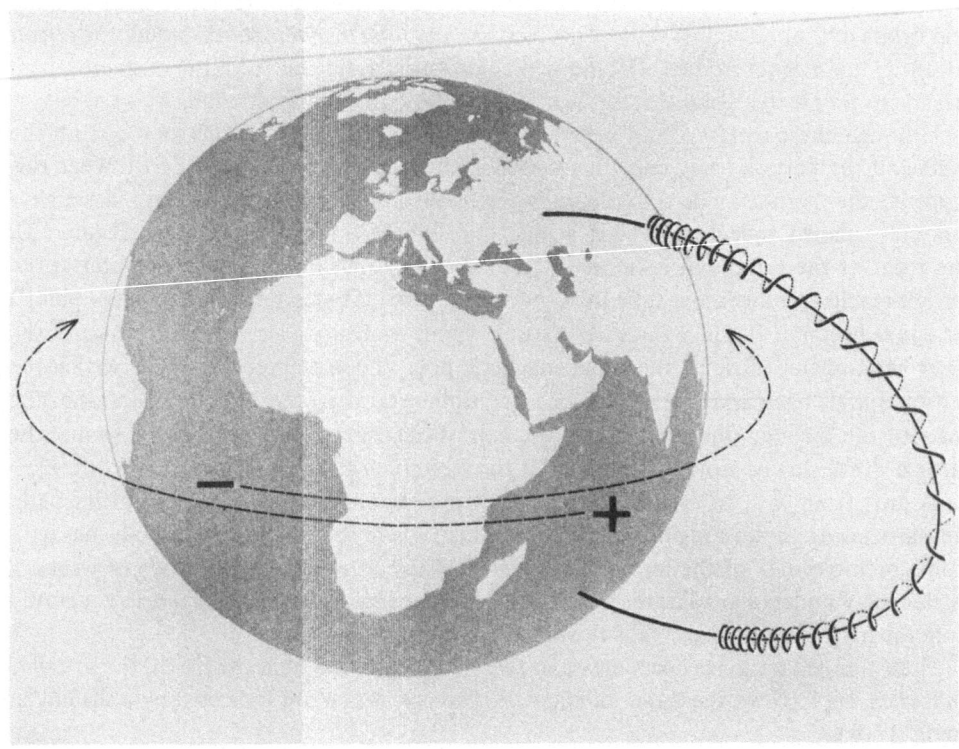

Fig. I-12. *Van Allen belts.* Charged particles which become trapped in the Earth's **magnetic** field spiral back and forth around a line of force. At the same time, they drift slowly around the Earth, eastward if they are negative, westward if positive. (Copyright © 1959 by Scientific American, Inc.)

Having finished with the explanation of Figure 1, and with it the general description of the atmosphere, we shall go to a more detailed study, which will deal mostly with the lower layers, particularly the troposphere. We shall consider the following aspects in different chapters: atmospheric chemistry, radiation, thermodynamics and vertical stability, cloud physics, atmospheric electricity and atmospheric dynamics.

Chapter I: Questions

Q1. Explain why atmospheric pressure always decreases with height.

Q2. It is possible to introduce visible vapour trails at different altitudes in the atmosphere. Below 100 km, these become distorted and break up into puffs, whereas above 100 km they simply diffuse until they disappear. Explain the difference.

Q3. Give a qualitative explanation of the shape of the temperature curve as a function of altitude, from ground to the beginning of the exosphere.

Q4. Why are there high concentrations of free electrons in the ionosphere? Why does this not happen (*i*) at higher altitudes, and (*ii*) at lower altitudes?

Q5. Why are broadcastings from distant stations better received at night than during the day?

Q6. Why do the auroras occur (*i*) at high latitudes, (*ii*) in a given range of altitudes?

Q7. What is solar wind?

Q8. Why does the solar wind modify the Earth's magnetic field?

Q9. What are the Van Allen belts?

Chapter I: Problems

(Any necessary constants not given in the statement of a problem will be found in the Table of Constants on pages x–xi)

P1. Knowing that the Earth's radius measures 6370 km and assuming that the pressure is everywhere equal to 1 atm at the ground, compute the total mass of the atmosphere. Approximate the value of gravity by an average of $9.8 \, \text{m/s}^2$.

P2. Assume that the average temperature between 1000 and 900 mb in the atmosphere is 280 K. What is the thickness of the layer comprised between these two isobaric surfaces?

P3. If the tropopause is at 150 mb and the stratopause at 1 mb,
(a) Calculate the total mass per unit cross section of the stratosphere.
(b) How thick would the stratosphere be if brought to standard temperature (273 K) and pressure (1 atm)?
Assume that the value of gravity is constant and equal to $9.7 \, \text{m/s}^2$.

P4. (a) Assuming that the Earth is spherical, with mass centre at its centre, derive a formula for the decrease in acceleration of gravity g with increasing height over the surface.
(b) What is the mass of the Earth?

P5. Assume that at 1000 km of altitude and above the atmosphere is in approximate diffusive equilibrium, with a constant temperature of 1500 K, and that the number concentration of molecules of atomic oxygen O and of atomic hydrogen H at 1000 km is $n_O = 10^6 \, \text{cm}^{-3}$ and $n_H = 10^4 \, \text{cm}^{-3}$, respectively. Make an estimate of the altitude at which atomic hydrogen becomes more abundant than atomic oxygen. For the gravity, take an average constant value of $6.0 \, \text{m/s}^2$.

P6. Imagine an isothermal atmosphere composed of helium and neon, in diffusive equilibrium. At a certain level, $p_{Ne} = 4p_{He}$ (p_i = partial pressure of i). At what height above that level does helium become more abundant (in volume concentration) than neon? Assume an average gravity $g = 9.4 \, \text{m/s}^2$. The temperature is $T = 500 \, \text{K}$.

P7. Calculate the total number concentration of molecules for the following conditions:
(a) At the ground, with $p = 1000 \, \text{mb}$, $T = 20°\text{C}$.
(b) At 100 km of altitude, with $p = 10^{-3} \, \text{mb}$, $T = -50°\text{C}$.
(c) At 300 km of altitude, with $p = 3 \times 10^{-8} \, \text{mb}$, $T = 1500 \, \text{K}$.

P8. Pulses of 9 MHz sent by an ionosonde are reflected with a time lapse of 2 ms. What information, regarding electron density, do you derive from this?

P9. Considering only the presence of an E and an F region in the ionosphere, describe what you would expect to see on the screen of an ionosonde used with the first operation mode described in §4, as you slowly increase the frequency of the pulses emitted.

P10. Consider a region with uniform field of magnetic induction (or magnetic flux density) **B**. Assume that a proton is inserted
(a) at a velocity **v** perpendicular to the field;
(b) at a velocity **v** with components v_l along the field and v_t perpendicular to it.
Describe the shape of its subsequent trajectory in both cases.

II. Atmospheric Chemistry

Atmospheric chemistry deals with the following main questions:

(1) Composition of air and distribution of components.
(2) Sources and sinks of the different components, in the atmosphere and at its boundaries.
(3) Evolution in the atmosphere (chemical reactions).
(4) Cycles and budgets of main elements.

A detailed survey of these subjects would be beyond the scope of this book. We shall therefore limit ourselves to considering some general questions: the vertical distribution of water substance in the troposphere and stratosphere; a brief description of the cycles of sulfur, nitrogen and carbon compounds; the role of nitrogen dioxide and ozone in photochemical pollution; and the atmospheric aerosol.

Consideration will be restricted to the homosphere, and mainly to the troposphere. There are a number of problems that could be listed under the general denomination of 'atmospheric chemistry' but will be treated later; this includes two important fields: composition of the high atmosphere and ozone. These questions are closely related to radiation in the atmosphere, and are therefore left for Ch.III.

1. Composition of the Air

Atmospheric air can be considered as composed of:

(1) A mixture of gases which we call *dry air*, to be described below;
(2) Water substance in any of its three physical states; and
(3) Solid or liquid particles in suspension, called the *atmospheric aerosol*.

We shall start by considering the dry air.

Although there are a good many components in dry air, we can draw a sharp distinction between the more abundant *main components* and the *minor components* only present in trace amounts. Table 1 lists the air constituents. We may notice that, of the four main ones,

N_2, O_2 make up > 99% of dry air; N_2, O_2, Ar make up 99.97% of dry air; N_2, O_2, Ar, CO_2 make up > 99.997% of dry air

CO_2 is somewhat variable at the ground, as it is affected by any type of combustion (fires, industrial activities), photosynthesis and exchange with the oceans, but it is very

TABLE 1

Constituents of dry air

	Main constituents	Estimated residence time in atmosphere
	Molar (or volume) fraction	
N$_2$	0.7809	2 × 10^7 years
O$_2$	0.2095	
Ar	0.0093	
CO$_2$	0.00033	5–10 years
	Minor constituents	Permanent
Non variable	Concentration	
Ne	18 ppm (in volume)	
He	5 ppm	3 × 10^6 years
Kr	1 ppm	
Xe	0.09 ppm	
CH$_4$	1.5 ppm	3 years
CO	0.1 ppm	Semi- 0.35 years
H$_2$	0.5 ppm	permanent
N$_2$O	0.25 ppm	< 200 years
Variable	Typical concentration	
O$_3$	Up to 10 ppm in stratosphere	
	5–50 ppb (unpolluted air)	
	Up to 500 ppb in polluted air at ground	
H$_2$S	0.2 ppb (over land)	10 days
SO$_2$	0.2 ppb (over land)	5 days
NH$_3$	6 ppb (over land)	1–4 days
NO$_2$	1 ppb (over land)	2–8 days
	100 ppb in polluted air	
CH$_2$O	0–10 ppb	

constant above the surface layers. In fact, the dry air composition is remarkably constant throughout the whole homosphere, indicating that mixing processes are very efficient.

2. Minor Constituents

The minor components amount in total, as seen above and in the table, to less than 0.003% or 30 ppm*. They are, however, very important in the chemical picture of the atmosphere, even in problems of such practical importance as those related to pollution,

---

* The symbol ppm stands for the concentration unit 'parts per million'. This unit and ppb = parts per billion are commonly used when referring to trace gases. Unless we mention otherwise, 'parts' is here to be understood as 'parts in volume'.

or to the existence of the ozonosphere. Table 1 separates them in two groups: non-variable and variable. The reason will become clear in the following discussion.

Several classifications can be done of all the constituents of dry air, which illustrate the different roles played by them in the atmosphere. Let us consider them in turn.

(a) By abundance:

We have already mentioned that the four main constituents make up for more than 99.997% of the dry air, all four with concentrations > 300 ppm. The second group of the table, the non-variable minor constituents, are present in concentrations of 0.1 to 20 ppm. The variable components are all below 0.1 ppm (except in polluted air and stratospheric ozone). The distinction between the groups, regarding their abundance, is thus quite sharp.

(b) By variability:

All the main components are non-variable, and so are the first group of minor components in Table 1.

Variability is important in that it indicates something about the behaviour of the gas in the atmosphere. It is linked with abundance, with reactivity and with the residence time of the molecules in the atmosphere (see below). Thus CO_2 is non-variable in spite of localized sources of production, because the atmospheric reservoir of CO_2 is too large to allow appreciable fluctuations in concentration. Minor components of high reactivity, such as SO_2 or NO and NO_2, are variable because they react quickly, and the abundance is low.

We have not mentioned water vapour in the composition, as it will be considered apart. But of course it is a major component of atmospheric air, and the only one subject to considerable variations. These are linked with the processes of condensation and precipitation and, as many reactions of highly reactive gases may occur either in solution or with the help of water vapour, the evolution of these reactive, variable minor components of air is usually closely connected with the cycle of water.

(c) By chemical composition:

Obviously, this aspect is linked to the reactivity and other characteristics of the component gases. For instance, the complete inertness of noble gases makes them permanent and stable, even those at low concentrations. Most of the relevant chemistry in the atmosphere is concerned with compounds of sulfur and nitrogen. Carbon compounds, including some organic compounds, are also important. Other compounds, like halogen derivatives, play a smaller role.

(d) By residence time:

An important parameter of each gas for the atmospheric chemist is the *mean life* or *average residence time* τ. It is usually defined as

$$\tau = \frac{M}{F} \tag{1}$$

where M is the total average mass of the gas in the atmosphere, while F can be either the total average influx or outflux (which in time averages for the total atmosphere must be equal). $1/\tau$ is called the *rate of turnover*.

It is clear that if every molecule of gas stays in average τ in the atmosphere, it takes a time τ to effect a complete turnover; and as F is the influx, $F.\tau$ must be equal to the total mass M, thus justifying the previous definition.

The importance of τ is that it indicates how actively a given gas is going through a cycle. If M is small and the gas is very reactive, τ will be small and the gas concentration will be variable, because it does not have time to distribute homogeneously from the localized sources.

From the point of view of the residence time, the air components can be roughly classified in 3 categories:

(1) Permanent gases with very large τ, e.g. $\tau \sim 2$ million years for He.
(2) Semi-permanent gases, with τ from some months to years. This is the case of the four gases indicated in Table 1, as 'semi-permanent', which have several similarities in spite of their different chemical composition.
(3) Variable gases. With τ from days to weeks. These are the chemically active gases of the last group in the table. Their cycles are related to the water cycle; τ for water vapour is of the order of 10 days.

(e) By origin:

The following classification of origins, with examples of gases, covers the main cases:

Combustion – Natural – e.g. CO_2.
       – Anthropogenic – e.g. CO_2, SO_2, NO.
Biological Processes (bacterial activity, photosynthesis)
                    – e.g. CH_4, N_2O, H_2, NH_3, H_2S, NO.
Chemical Reactions in the Atmosphere – e.g. HCl.
Other origins, like volcanic activity, are of minor importance.

TABLE 2

Summary of important chemical species in atmospheric chemistry

S compounds		N compounds		C compounds		Others	
H_2S		NH_3, NH_4^+	R	CH_4	B NV	H_2	B NV
SO_2	R	N_2O	B NV	CO	B NV	O_3	
$SO_3, SO_4^=$		NO		CO_2	NV		
		NO_2	R				
		NO_3^-					

Frames = compounds involved in the same cycle.
    R = reactive; cycle linked to water cycle; short τ; variable.
    B = mainly biological.
NV = non-variable (semi-permanent).

3. Summary of Important Compounds in Atmospheric Chemistry

In Table 2, we have made a simplified classification of the most important compounds to be considered in atmospheric chemistry. No consideration is given here to N_2 and noble gases, for their inertness, or to O_2, although this gas takes part in oxidation reactions.

The frames in the table indicate the groups of species that are related to each other by chemical reactions, so that they must be considered together in a cycle. Thus, for instance, H_2S, SO_2 and SO_3 are three compounds related by oxidation reactions; when SO_3 has been formed, it will rapidly absorb water vapour to produce H_2SO_4, which may combine with basic substances in the atmospheric aerosol to give sulfates; this is indicated by $SO_4^=$. This group, that of NO, NO_2, NO_3^- and that of NH_3, NH_4^+, contain the reactive gases, of short residence times, high variability, low concentration and related to the water cycle.

TABLE 3

Sources and sinks

	Sources	Compounds	Sinks
S compounds	Biological (bacterial decomposition of organic matter)	→ H_2S	
	Combustion	→ SO_2	→ Rainout, washout
	Sea spray	→ $SO_4^=$	→ Plant intake
N compounds	Biological	→ NH_3, NH_4^+	→ Rainout, washout
	Bacteria in soil	→ N_2O	→ { Bacteria in soil, Decomposition in stratosphere
	Biological, combustion	→ NO, NO_2, NO_3	→ Rainout, washout
C compounds	Biological (paddy fields, marshes) / Natural gas	→ CH_4 →	{ Oxidized at ground (?), Decomposition in stratosphere (10%)
	Photosynthesis by microorganisms at sea and by the terrestrial biosphere	→ CO →	{ Bacteria in soil, Reaction in stratosphere
	Combustion / Breathing / Bacteria (decomposition of organic matter)	→ CO_2 →	{ Photosynthesis, $CaCO_3$ in oceans
Others	Biological (decomposition of organic matter)	→ H_2 →	Bacteria

25

Apart from CO_2, the four gases framed by themselves in the table, N_2O, CH_4, CO and H_2, present similarities, as was mentioned above: they have little reactivity, are relatively abundant in the atmosphere, they are therefore little variable and have longer residence times, and their cycles are mainly biological (bacterial activity), at least regarding their origin.

CO_2 is a separate case, being one of the main constituents; it has its own separate cycle.

O_3 must be considered apart, in that it takes a part in the cycle of NO, NO_2, and probably also in that of the sulfur compounds, as an oxidant.

Tables 1 and 2 are completed by Table 3, which summarizes the main sources and sinks of the different compounds. We shall further develop the consideration of their chemistry and cycles in §§5–9, but we shall first give some attention to an essential major constituent of air, whose peculiar properties have the most important consequences in the behaviour of the atmosphere: water substance.

4. Water Substance

The cycle of water substance is linked to very important processes in the atmosphere:

(1) Condensation and evaporation of water have important consequences for the thermodynamics of atmospheric processes and for the vertical stability of the atmosphere. These aspects will be considered in Ch.IV.

(2) Water substance is, of course, essential in the formation of clouds and the development of precipitation. This will be the subject of Ch.V.

(3) The water cycle is important in the cleansing of the atmosphere by two mechanisms:

Rainout – This is the name given to the processes of removal of substances occurring within the clouds. The main process is the participation of hygroscopic aerosol particles in the formation of cloud droplets ('cloud condensation nuclei': see Ch.V); many of these cloud droplets become part of raindrops or other precipitation elements and are thus carried down to the ground. Other processes of minor importance, like capturing of small particles by the droplets, due to Brownian motion (cf. §10), may also contribute in rainout.

Washout – This is the name given to the elimination of gases by dissolution and of aerosol particles through capture by falling water drops. Washout includes the processes occurring below the clouds.

(4) Water substance is directly involved in atmospheric chemistry either as water vapour participating in chemical reactions or through reactions of other substances occurring in aqueous solution.

(5) Both water vapour and clouds play important roles in radiation transfers through the atmosphere. This will be treated in Ch.III.

Water substance can also be used as an indicator of circulation in the stratosphere (see below). The isotopic composition of water (content in HDO^{16} or in H_2O^{18}) can be used as a tracer in the study of certain processes.

Consideration of water substance in the atmosphere is therefore of the utmost importance. For brevity's sake, the hydrological cycle will be omitted in this book[*] and in this chapter we shall only consider the vertical distribution of water vapour in the troposphere and stratosphere.

The concentration of water vapour is very variable and in spite of the large total mass contained in the atmosphere, its residence time is very short; it can be estimated as ~ 10 days. But there is, for every temperature, an upper limit for the concentration of water vapour in air; this is given by the partial pressure that saturates it. By definition, this *saturation vapour pressure* is the partial pressure at which water vapour can coexist in equilibrium with liquid water. It is practically independent of the presence and pressure of dry air. It can be derived, thermodynamically, that the dependence of saturation vapour pressure with temperature is given by the expression

$$\frac{\mathrm{d}\ln p_s}{\mathrm{d}T} = \frac{\ell_v M_v}{RT^2} \tag{2}$$

called the Clausius–Clapeyron equation. Here p_s = saturation vapour pressure, T = absolute temperature, ℓ_v = latent heat of vaporization, M_v = molecular weight of water vapour, R = gas constant. ℓ_v varies with T, but not much; if in first approximation we consider it as a constant, we can integrate (2) to:

$$p_s = \mathrm{const} \times \mathrm{e}^{-\ell_v M_v / RT} \tag{3}$$

which indicates a very rapid increase of p_s with temperature. The actual curve $p_s = f(T)$ is shown in the graph of Figure 1. An immediate consequence of these properties is that only the lower, warmer, layers of the troposphere can contain high concentrations of water vapour. In cold regions and in the higher layers (which are always cold), the water vapour is sharply limited by the saturation value. Thus, we see in the graph that the partial pressure of water vapour can reach up to about 32 mb in air at $25°$C, but only to 1.2 mb at $-20°$C. When humid air becomes colder (for instance, due to adiabatic ascent, as described in Ch.IV), water vapour condenses, giving rise to the formation of clouds. Notice that below $0°$C water vapour can condense into water (supercooled water) or into ice; Figure 1 shows both equilibrium curves (supercooled water–water vapour and ice–water vapour) below $0°$C.

Let us consider now the vertical distribution of water vapour in the troposphere and stratosphere in middle latitudes. The measurements in the stratosphere can be performed by different methods, including determination of dew point (see Ch.IV, §4) in flights, spectroscopy and dew point radiosondes. Although these measurements have been the subject of some controversy, it seems now well established that very low humidities exist in the lower stratosphere above middle latitudes. The situation is summarized schematically in Figure 2, where an approximate curve (full line) suggested by a number of cases has been drawn, indicating the variation of humidity with height. The humidity variable is here the *mass mixing ratio*, defined as the mass (in g) of water vapour per unit mass (kg) of dry air; the advantage of this stoichiometric variable being that it is not dependent on the pressure changes or on any process not involving condensation. Thus, if there was an active exchange of air through the tropopause (whose level is indicated), the mixing

---

[*] The book by Barry and Chorley (see Bibliography), for instance, can be consulted on this subject.

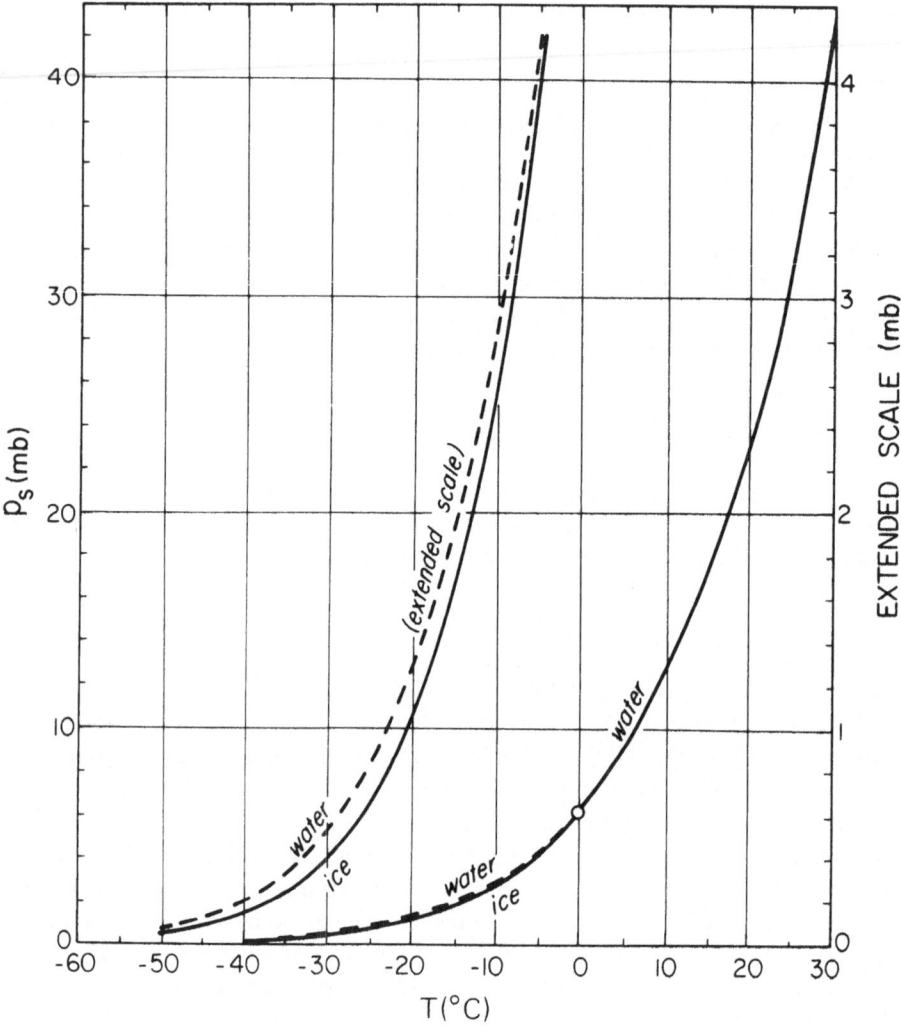

Fig. II-1. *Saturation vapour pressure of ice and water.* p_s = pressure; T = temperature.

ratio should keep the same value above the tropopause, because in this region the temperature does not decrease any longer with height and therefore no further condensation can occur. The temperature curve has also been indicated (dotted line) for the region of interest. The interesting feature of this graph is the extremely low values of humidity in the lower stratosphere, with the sharp decrease continued above the tropopause. A third curve (dashed) gives the humidity in terms of the frost point, which can be defined as the temperature at which the humidity corresponds to saturation over ice, for that local pressure: see Ch.IV, §4. We see that the humidity decreases from a mixing ratio of about 0.012 g/kg (or a frost point of $-66°C$) at the tropopause to about 0.0025 g/kg (or below $-80°C$) a few kilometers higher up. This is a surprising result, indicating that the air in the lower stratosphere cannot have come from the troposphere by exchange through the tropopause.

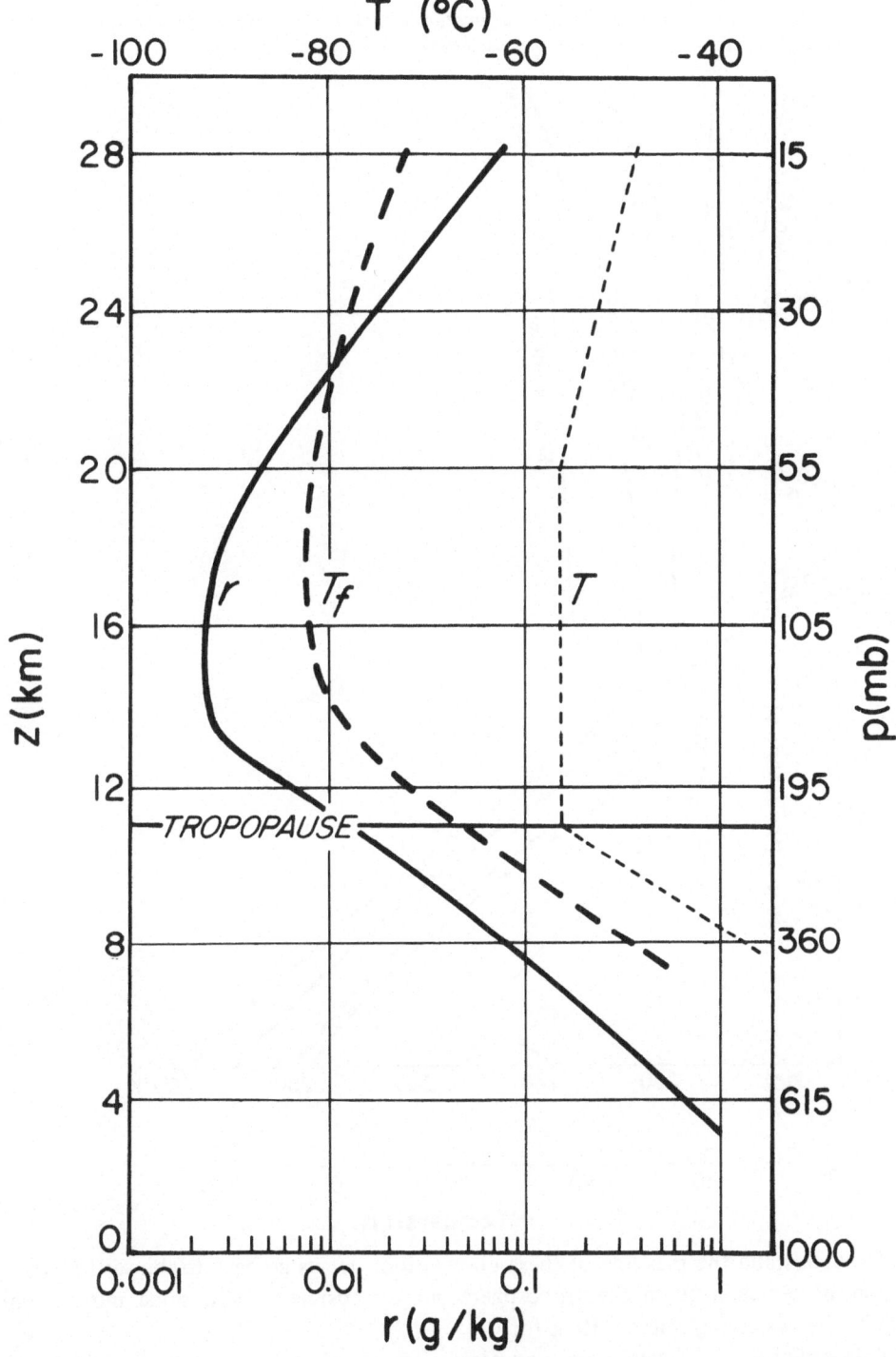

Fig. II-2. *Humidity as a function of altitude at middle latitudes.* z = altitude; T = temperature; p = pressure; r = mass mixing ratio; T_f = frost point. The curves and figures must be taken as an illustrative example rather than general averages.

Fig. II-3. *Average curves of temperature as a function of height at different seasons and latitudes.* On individual days, the temperature may vary considerably from these mean values.

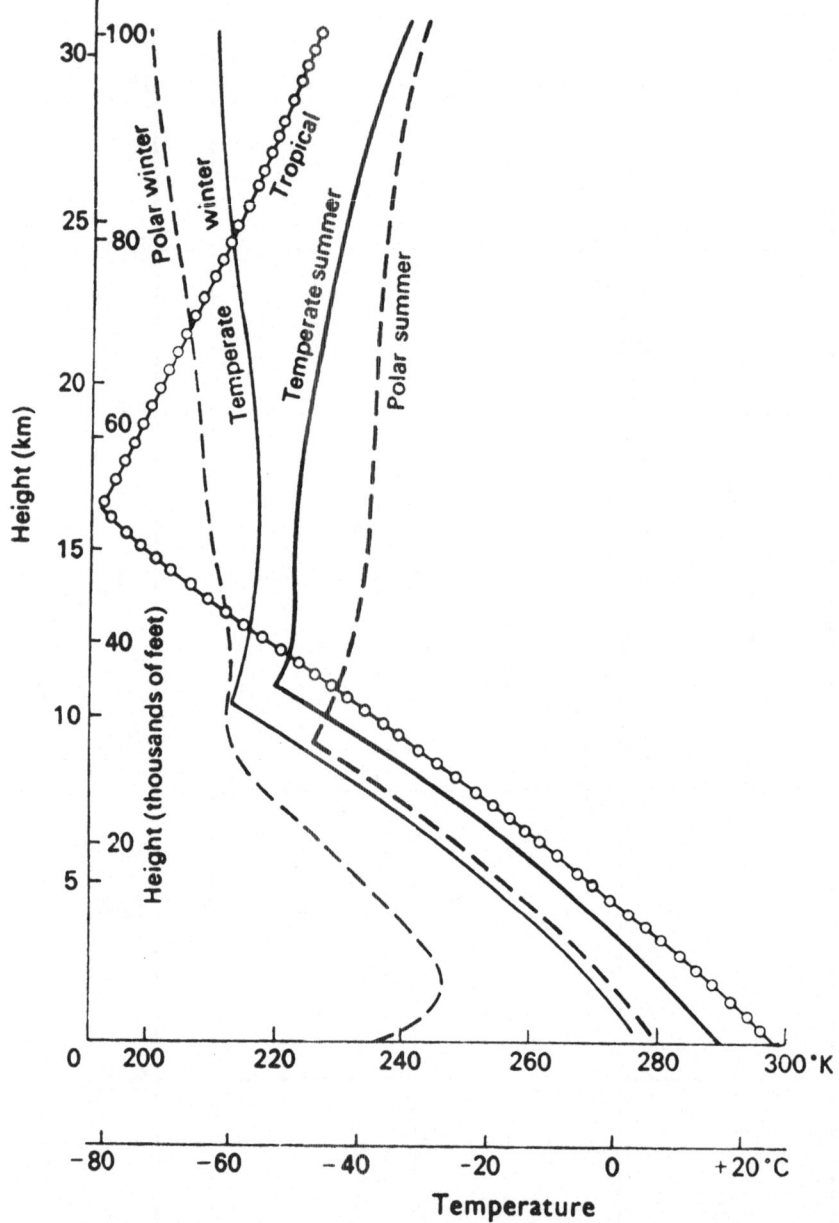

To understand the meaning of these observations, let us go back to the vertical distribution of temperature in the troposphere and stratosphere, with some more detail. Figure 3 shows average distributions for

    Polar regions   — winter and summer
    Temperate regions — winter and summer
    Tropical regions   — (no difference between winter and summer)

The graph shows that the tropopause (sudden change in temperature trend) is much higher in the tropics, and therefore, in spite of starting at higher surface temperatures, the temperature reaches much lower values — reaches, in fact, a temperature of − 80°C at about 17 km. Also, the polar stratosphere in summer has the highest temperature. Regarding water vapour, what this indicates is that the stratosphere air at temperate latitudes must have come from the troposphere through the tropical tropopause, in order to explain its low water vapour content; in crossing that region, water vapour will condense into ice crystals, which will fall out and leave the air with a saturation vapour pressure corresponding to ice at − 80°C. There is, therefore, a meridional circulation in the stratosphere, going from the tropics towards higher latitudes. This is shown schematically in Figure 4.

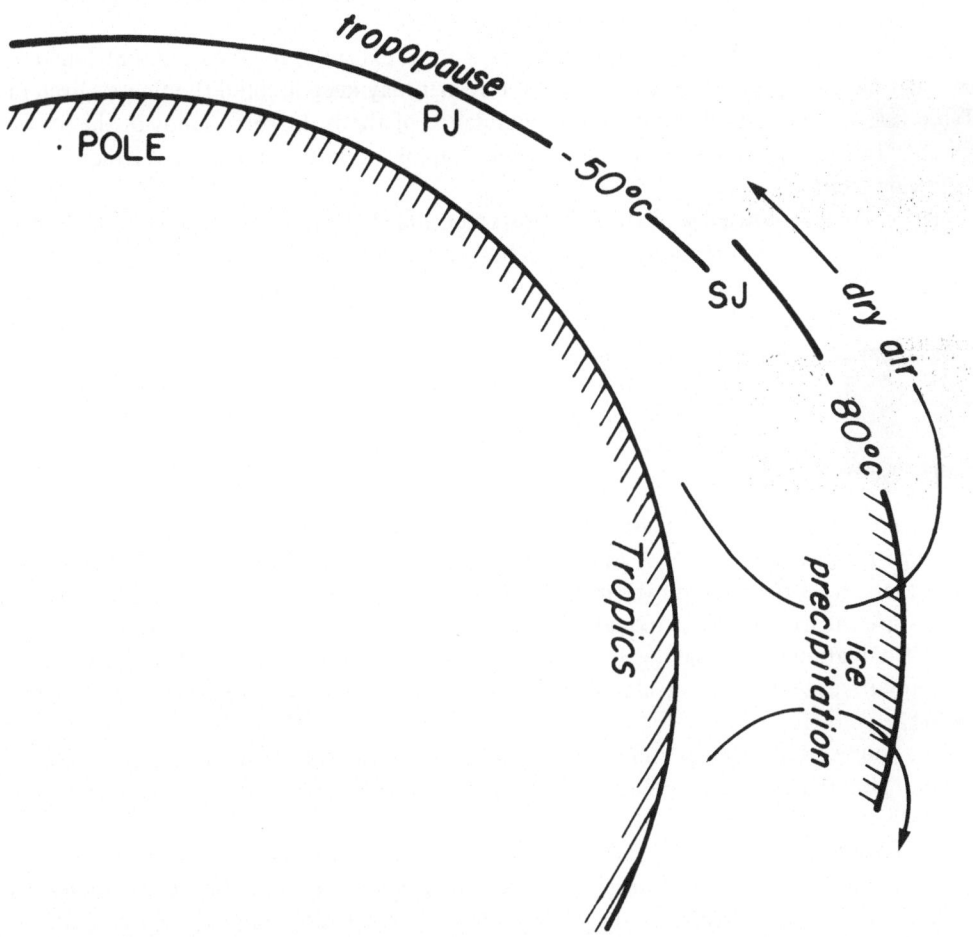

Fig. II-4. *Meridional circulation in the stratosphere*. In order to explain the low humidity, air in the lower stratosphere is assumed to come from the troposphere through the tropical tropopause, whose temperature is very low. At that stage, water vapour condenses to ice crystals and precipitates, as indicated schematically in this figure (cf. Figure III, 10). The two interruptions in the tropopause correspond to the subtropical jet (SJ) and the polar jet (PJ) (cf. Ch. VII, § 10).

5. Cycles of the Main Elements

The chemical substances playing a role in atmospheric chemistry can be conveniently grouped, according to their chemical composition, as the derivatives from different elements. The three main groups are then the sulfur compounds, the nitrogen compounds and the carbon compounds. Other groups, like halogens, are very scarce and play a minor role, or else are too inert to become involved in chemical reactions (noble gases).

The complete understanding of the atmospheric part of the cycle of each one of these groups would imply a knowledge of:
1. The sources of the different compounds
2. The chemical transformations undergone in the atmosphere
3. The sinks for the different compounds
4. The distribution and abundance of the different compounds in the atmosphere
5. The budget of transfers and transformations.

Here we shall omit the study of the last two points, except as for the figures included in considering sources and sinks. However, the reader may keep in mind the figures given in Table 1 as a general indication of the abundance of the main compounds in the atmosphere. Tables 2 and 3 will also find general application as summaries related to the following sections.

The complete knowledge of the five points mentioned above is a goal which is obviously not easy to attain. Estimates for the global budget figures over the Earth are, by necessity, crude and subject to revisions. The chemical transformations involved are frequently badly understood. Nevertheless, an understanding of these cycles, however incomplete or provisional, is of primary importance.

6. The Sulfur Compounds

The main sulfur compounds can be classified in three stages of oxidation of the sulfur element. Hydrogen sulfide H_2S and organic derivatives, such as dimethylsulfide $(CH_3)_2S$ and others, have the sulfur in the most reduced state (valency -2). Sulfur dioxide SO_2 is in an intermediate state of oxidation (valency $+4$) and sulfur trioxide SO_3 is in the highest oxidation state (valency $+6$). SO_3 is related to sulfuric acid H_2SO_4 (of which it is the anhydride) and sulfates (salts of the sulfuric acid), which we shall indicate generically by $SO_4^=$.

The sources and sinks of sulfur compounds in the atmosphere are described by a convenient pictorial sketch in Figure 5. The figure is largely self-explanatory, and we shall only add some comments.

The most important natural sources are the bacterial decomposition of organic material from flat coastal surfaces, swamps, etc., which gives H_2S and other reduced compounds to the air, and sea spray carried by the wind, which leaves particles containing several salts — mainly NaCl, but also including sulfates. Of comparable importance are the industrial activities, by which the burning of S-containing fuels introduces SO_2 into the atmosphere; thus the anthropogenic influence on this cycle is a very important one. The main sink, for SO_2 and sulfates, is the rain that carries these compounds back to the ground (washout and rainout); this is indicated in the figure as 'wet deposition'. Absorption of SO_2 by

Fig. II-5. *The sulfur cycle.* Sources and sinks of atmospheric sulfur compounds. The figures are given in Tg of S/year (1 Tg = 1 teragram = 10^{12} grams – 1 million metric tons), and are based on estimates by Granat et al. (see Bibliography). 'Tg of S' means that the transfers are calculated in mass of sulfur, whatever the compound involved. The main compounds transferred are indicated in each case. The left side of the sketch refers to sources and sinks over land, and the right one to those over oceans. A transport of 2 Tg/y from land to ocean, through the atmosphere, is necessary to make the figures consistent.

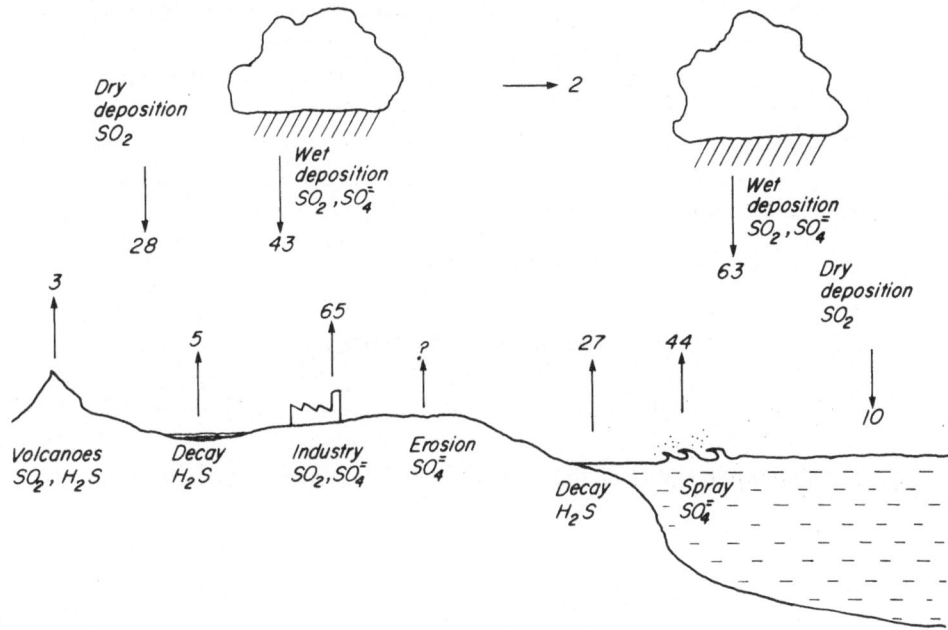

plants and deposition of particles containing sulfates contribute to the elimination of sulfur compounds from the atmosphere; this is called 'dry deposition'.

All three sulfur compounds – H_2S, SO_2 and sulfates – are here being considered together. Budget considerations make it obvious that all are part of the same cycle, because they are related by chemical transformations in the atmosphere. For instance, it would be very difficult to explain how large amounts of H_2S and other reduced compounds introduced into the atmosphere could disappear as such; it is obvious that H_2S is mostly oxidized by some mechanism, before being eliminated in a sink, as SO_2 or sulfates. Let us consider now the chemical reactions.

(i) *Oxidation of H_2S to SO_2.* This is, in fact, an unsolved problem. Oxidation in the gas phase may happen through complex reactions involving atomic O and radicals. In solution (cloud droplets), H_2S can be oxidized by ozone:

$$H_2S + O_3 = SO_2 + H_2O \tag{4}$$

(ii) *Oxidation of SO_2 to SO_3.* This can occur in a number of ways, but two seem to be the main processes.

a) In gas phase, by photochemical action. The oxidation occurs ultimately by oxygen from the air, but in order for the reaction to occur fast enough, the action of solar radiation is necessary, perhaps to bring the SO_2 molecule to a state of higher energy that can react more easily ('excitation'; see Ch.III). In spite of numerous studies, this is still a reaction not fully understood.

b) In liquid phase (i.e. in water droplets) SO_2 may become oxidized more readily. CO_2 and NH_3, always present in the air, dissolve as well as the SO_2, and play a role in this process. The SO_2 passes to the state of $SO_4^=$ (sulfate ion) and if the main cation in solution is NH_4^+ (from the dissolved NH_3), the droplet will eventually evaporate to a residue of ammonium sulfate $(NH_4)_2SO_4$.

(iii) *Formation of Sulfates*. We have just considered above one possible mechanism for the formation of ammonium sulfate. If SO_2 is oxidized to SO_3 in gas phase, some transformations follow readily. SO_3 absorbs immediately water vapour from the air and forms a droplet of H_2SO_4 in solution. This can absorb NH_3 and produce $(NH_4)_2SO_4$ as before. Also, if H_2SO_4 appears in cloud droplets, either by capture or by formation in the droplet itself, and if this droplet contains NaCl, by subsequent evaporation the following reaction may take place:

$$H_2SO_4 + 2\,NaCl = Na_2SO_4 + 2\,HCl \tag{5}$$

where HCl evaporates into the air, and sodium sulfate remains as a residue. Other similar reactions can also occur with other cations.

We must remember that sulfate-containing particles are also produced directly from sea spray and from land erosion.

Sodium, ammonium and calcium sulfates are abundant in aerosol particles.

7. The Nitrogen Compounds

The cycle of nitrogen in nature can be considered separately for four different groups of compounds:

   a) Gaseous molecular nitrogen N_2 and nitrous oxide N_2O.
   b) Ammonia NH_3 and derivatives.
   c) Oxides other than N_2O, and their related compounds. We shall indicate this group by NO_x.
   d) Organic compounds containing nitrogen.

   a) N_2 is, of course, the most abundant chemical species in the atmosphere, and its participation in the cycle is minimal when compared with the enormous reservoir constituted by the atmosphere. This participation is mainly limited to its fixation in the soil or in the oceans by microorganisms that transform it into organic compounds, and its return to the atmosphere by the action of bacteriae that reduce nitrates (NO_3^-) and nitrites (NO_2^-). In these reductions N_2O is also produced, and this is its main source. The sinks of N_2O are not well known, except for photochemical decomposition in the stratosphere:

$$N_2O + h\nu = N_2 + O \tag{6}$$

where $h\nu$ indicates radiation of a certain frequency ν (see Ch.III). But the main sinks are probably of biological nature. In any case, this rather abundant and stable gas (cf. Table 1) does not appear linked to the cycle of other nitrogen compounds.

   b) Ammonia is a gas that contains nitrogen in its most reduced form (valency -3).

Ammonia compounds are reactive and variable. Their cycle is linked with that of water. NH_3 dissolves readily in water, with production of ammonium hydroxide, which, in turn, goes into a dissociation equilibrium:

$$NH_3 + H_2O \rightleftarrows NH_4OH \rightleftarrows NH_4^+ + OH^- \tag{7}$$

The hydroxyl ions OH^- may combine with hydrogen ions H^+ (giving H_2O), if these appear, due to the presence of an acid. In that case, because the OH^- disappears from the equilibrium and due to the law of mass action, (7) becomes displaced toward the right, and more ammonia can dissolve in the system. For instance, the presence of sulfuric acid, which dissociates to $SO_4^= + 2H^+$, will lead to a solution containing NH_4^+ and $SO_4^=$; when a droplet of this solution evaporates, a particle of ammonium sulfate $(NH_4)_2SO_4$ will remain. Indeed, this salt is one of the most abundant components of the atmospheric aerosol. Similarly, carbon dioxide CO_2 which is always present in the air will help the ammonia to dissolve in the cloud droplets, giving a solution where the predominant ions will be NH_4^+ and HCO_3^- (i.e. a solution of ammonium bicarbonate).

The main source of NH_3 is the biological decomposition of organic matter. Once in the atmosphere, it may undergo chemical transformations such as those indicated above, and return to the ground or the ocean by dry or by wet deposition*. Only a very small fraction of the terrestrial source can be attributed to combustion of fuels; the anthropogenic input is here negligible.

This cycle, which is described by Figure 6, appears independent of that of NO_x, except for a small fraction of NH_3 undergoing decomposition by the radical OH to produce nitrogen oxides, and thus providing a small link between the two cycles.

c) In the generic formula NO_x we include the two main oxides present in the atmosphere (other than N_2O) – nitric oxide NO and nitrogen dioxide NO_2 – and a number of other species which play important intermediate roles in a rather complicated set of chemical reactions but are of little abundance. These would include N_2O_3 and the corresponding nitrous acid HNO_2 of which it is the anhydride, and N_2O_5, which by hydration gives the nitric acid HNO_3. In the course of these reactions, salts of these acids, nitrites (NO_2^-) and nitrates (NO_3^-), may also be produced, contributing significantly to the atmospheric aerosol. Nitrates have the nitrogen at the highest oxidation stage (valency + 5) and are quite stable; thus they represent the final state of a series of oxidations. Examples of a few important reactions are schematically indicated by:

$$NO \xrightarrow{\text{oxidation by } O_3} NO_2 \xrightarrow{\text{oxidation by } O_3} NO_3^-$$

$$NO_2 \xrightarrow[\substack{\text{decomposition} \\ \text{(see §9)}}]{\text{photochemical}} NO + O$$

---

* Similarly to what was indicated for sulfur compounds, 'dry deposition' includes here slow deposition of solid aerosol particles (with ammonium salts) and also direct elimination of gaseous NH_3 by absorption at the ground or oceans. 'Wet deposition' includes the processes of washout and rainout described before.

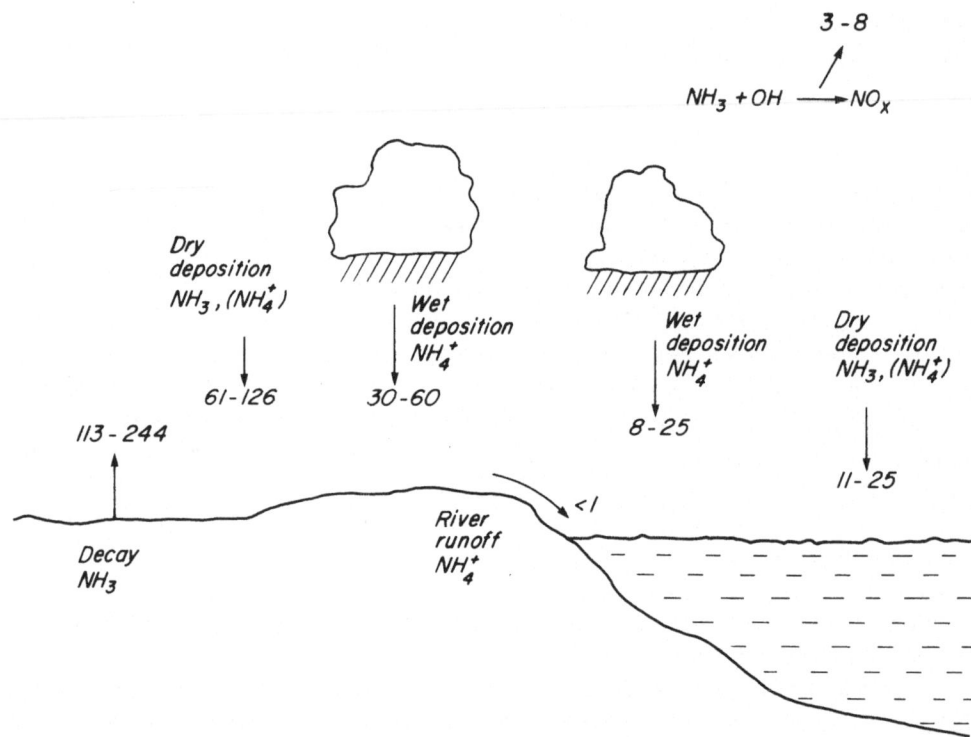

Fig. II-6. *The ammonia cycle.* Sources and sinks of atmospheric ammonia and ammonium compounds. The estimated ranges of transfers are given in Tg of N/year (1 Tg = 10^{12} g; all compounds expressed in mass of contained nitrogen). The main compounds involved in the transfer are indicated in each case; NH_4^+ means ammonium salts; brackets indicate minor contributions. The left side of the sketch refers to land, and the right one to oceans. The figures are based on estimates by Söderlund and Svensson (see Bibliography).

The main source of NO_x is bacterial decomposition of nitrates from the soil. Industrial combustion of fossil fuels is an additional appreciable source. A minor contribution is the oxidation of NH_3 by OH, which links this cycle to that of ammonium compounds. Figure 7 summarizes the cycle.

This is a cycle of very reactive and variable compounds (cf. Table 1), linked to the cycle of water substance through the participation of H_2O in the reactions, as well as through the processes of wet deposition.

d) Organic N-containing compounds are extremely important in the soil and oceans, but play a less important and little-understood role in the atmosphere. Turnovers of the order of several tens of Tg/year are probably involved.

8. The Carbon Compounds

The carbon compounds must be considered in separate cycles, which are not interconnected:

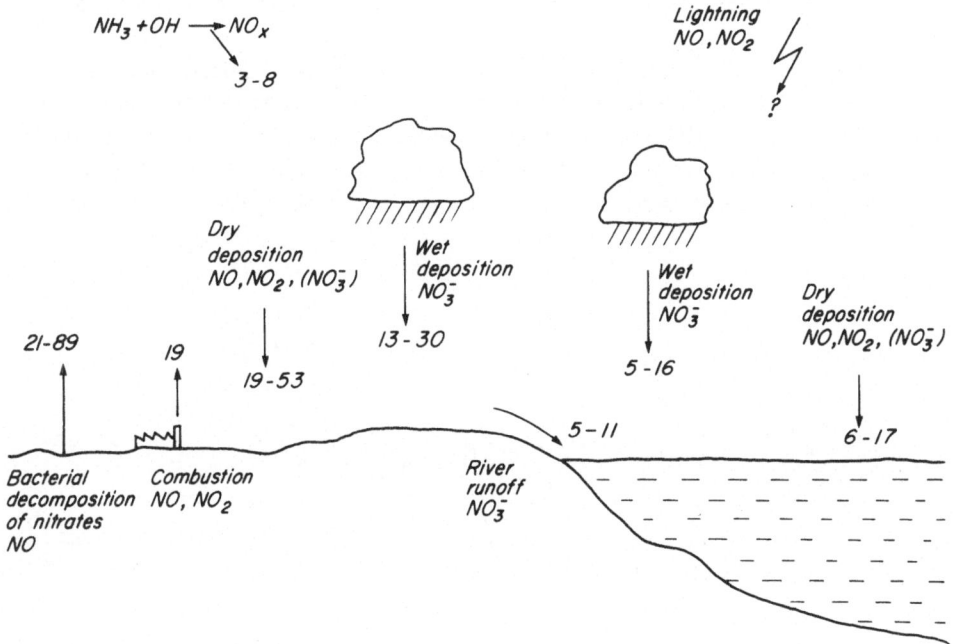

Fig. II-7. *The NO_x cycle.* Sources and sinks of NO, NO_2 and related compounds (see text). The estimated ranges of transfers are given in Tg of N/year (1 Tg = 10^{12} g; all compounds expressed in mass of contained nitrogen). The left side of the sketch refers to land, and the right one to oceans. The main compounds involved in the transfer are indicated in each case; NO_3^- stands for nitrates; brackets indicate minor contributions. The figures are based on estimates by Söderlund and Svensson (see Bibliography).

a) Carbon dioxide CO_2 c) Methane CH_4
b) Carbon monoxide CO   d) Others

a) CO_2 is one of the major constituents of atmospheric air (cf. Table 1). Its main sources are the combustion of C-containing substances, either in natural processes (forest fires) or in man's activities (combustion of fossil fuels, etc.), the animal breathing, and the decay of organic material. Its main sinks are plant photosynthesis of carbohydrates

$$nCO_2 + nH_2O \xrightarrow{\text{light}} (CH_2O)_n + nO_2$$

and solubility in the oceans, leading to slow penetration to deep layers and formation of calcium carbonate ($CaCO_3$) sediments.

CO_2 dissolves readily in water, producing the weak carbonic acid, which in turn goes into a dissociation equilibrium:

$$CO_2 + H_2O \rightleftarrows H_2CO_3 \rightleftarrows H^+ + HCO_3^- \qquad \qquad H^+ + CO_3^= \qquad (8)$$

37

Due to the law of mass action, alkalinity (low concentration of H^+) displaces these equilibria toward the right. The equilibrium between CO_2 in the atmosphere and the CO_2 dissolved in the upper layers of the ocean is therefore closely linked to the pH in these waters.

Increasing production of CO_2 due to industrial expansion and decreasing forest cover (a major sink for CO_2) have led to a regular increase of its content in the atmosphere, as illustrated in Figure 8. The concentration is presently about 330 ppm, but this is increasing at a rate close to 1 ppm per year. This situation gives cause for concern, because of the role of this gas in the transfer of radiation through the atmosphere, to be considered in Ch.III.

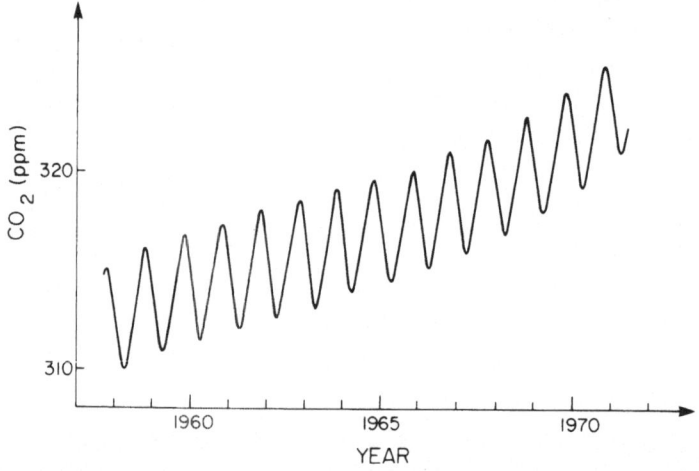

Fig. 11-8. *Variation of the CO_2 concentration.* The curve is based on results obtained in Mauna Loa, Hawaii (3400 m altitude). It shows a marked seasonal variation, attributable to biological activity of vegetation in the Northern Hemisphere, and a regular increase of the yearly average of 0.7–0.8 ppm/year.

b) Although CO can be oxidized to CO_2, this requires strong chemical conditions not met in the atmosphere, so that the two gases do not interact. CO has a cycle of its own, dominated by biological controls: it seems to be mainly produced by microorganisms at the surface of the ocean, and the main sink is probably by bacterial consumption in the soil. An additional source is by incomplete combustion, and additional sinks are photosynthesis and destruction by chemical reaction with the radical OH in the stratosphere.

CO is a relatively abundant and stable gas, as indicated in Table 1. However, it becomes extremely variable in cities and in forests (effect of localized sinks and sources).

c) Like CO, CH_4 is a relatively abundant and stable gas controlled by a biological cycle. It is produced by decay in marshes, swamps, paddy fields, etc. The sinks are not known; it may be oxidized or destroyed biologically. A minor fraction is destroyed by OH in the stratosphere.

d) Other C compounds include a number of organic substances, like terpenes, other hydrocarbons, formaldehyde, etc. They are of minor importance and will not be considered.

9. Photochemical Pollution

As a chemical problem of particular interest, we shall briefly mention now the basic facts underlying the formation of 'smog', in cities of high levels of pollution.

The starting point is a photochemical reaction, i.e. a reaction that occurs under the influence of light. We shall go into more details with respect to this type of reaction in Ch.III. Here it may suffice to mention that if NO_2 absorbs light of wavelength smaller than 385 nm, it becomes dissociated into NO and atomic O. In the air, this is quickly followed by two more reactions as indicated below:

$$NO_2 + \text{light of } \lambda < 385 \text{ nm} = NO + O$$

$$O + O_2 + M = O_3 + M$$

$$\underline{O_3 + NO = O_2 + NO_2}$$

$$\text{addition: no chemical reaction} \tag{9}$$

(M = molecules not reacting, but acting as a 'third body' to remove energy released in the chemical reaction)

The atomic oxygen immediately forms ozone with the oxygen of the air and the ozone re-oxidizes the nitric oxide NO to the original nitrogen dioxide NO_2. Adding the three reactions it is seen that no net chemical reaction occurs. In a pure atmosphere, therefore, this has the only effect of maintaining in the air a certain concentration of O and O_3. These concentrations are controlled by the velocity of the three reactions; as the second reaction is extremely fast, the concentration of atomic oxygen is kept at an extremely low level. The third reaction is not so fast, and the ozone concentration will have higher values (of the order of 0.005–0.05 ppm).

If, on the other hand, the atmosphere becomes polluted by substances easily oxidized by ozone, the corresponding oxidation reactions will compete for the ozone with the third reaction of the previous cycle, thus diverting ozone from it. This happens with hydrocarbons (compounds of C and H), particularly the *olefins*, which are hydrocarbons containing highly reactive double bonds such as $CH_2 = CH_2$ (ethylene), $CH_2 = CH–CH_3$ (propylene), etc. The reactions of ozone with hydrocarbons and subsequent reactions of the products are extremely complex, and among the final products some are acutely irritant to eyes and skin and show phytotoxicity. These noxious effects are obtained from the exhaust gases of cars, which provide at the same time abnormally high concentrations of NO_2 in the air (up to the order of 0.1 ppm), which in turn produces high concentrations of O_3 (up to the order of 0.5 ppm) by the above reactions, and high concentrations of hydrocarbons from the incomplete combustion of gasoline.

The three compounds mainly responsible for eye irritation and phytotoxicity of photochemical smog are:

Formaldehyde $\qquad\qquad CH_2 = O$
Acrolein $\qquad\qquad\qquad CH_2 = CH–CH = O$
Peroxyacetylnitrate (PAN) $CH_3–CO–O–O–NO_2$.

10. The Atmospheric Aerosol

Atmospheric air contains a great many solid and liquid particles in suspension, which constitute what we call the *atmospheric aerosol*.

The origin and chemical composition of these particles may be summarized in the following list of sources:

1. Combustion: forest fires, industrial combustions. The particles may have various salts, carbon, soot.
2. Gas phase reactions, including photochemical. We have seen examples in §§6 and 7, with the formation of sulfates and nitrates.
3. Dispersion of solids. Chemical reactions in the ground followed by water erosion and wind erosion can result in the introduction of particles from mineral rocks into the air: silicates, sodium, potassium and calcium salts, etc.
4. Dispersion of solutions. The bursting of tiny bubbles in the sea introduces particles in the air, essentially with the composition of sea water.
5. Volcanoes.

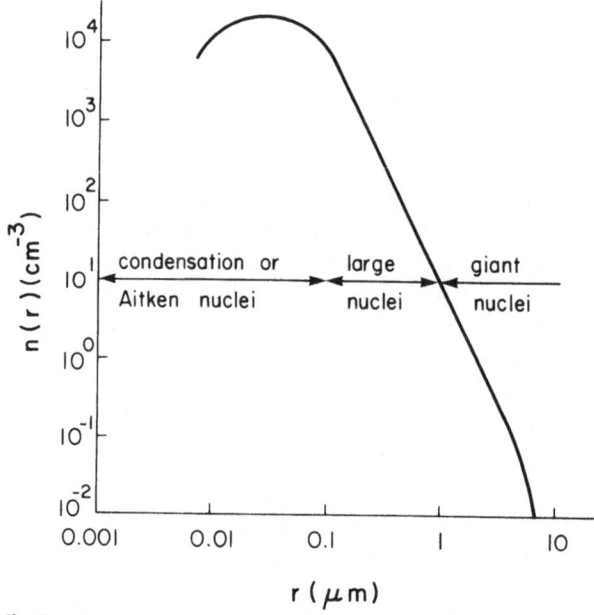

Fig. II-9. *Size distribution in continental aerosol*. The figure gives a typical plot of the distribution function over $\log r$: $n(r) = -\,dN/d \log r$, where r is the spherically-equivalent radius and N is the number concentration of particles with radius $> r$. (Based on results by C. Junge.)

TABLE 4
Typical concentrations of particles at ground (cm^{-3})
(Variability: ± one order of magnitude)

Over ocean	10^3
Over country	10^4
City	10^5

Table 4 and Figure 9 give an indication of the abundance of these particles at the ground in different locations and of the typical size distribution in continental air, respectively. The abundance decreases rapidly with altitude, reaching concentrations of the order of 10^2 in the high troposphere. Regarding the size distribution, it can be seen in the graph that there is, through most of the range from $0.1\,\mu m$ to $10\,\mu m$, a continuous decrease in concentration that obeys the so-called Junge's law:

$$n(r) = -\frac{dN}{d\log r} = \frac{A}{r^3} \tag{10}$$

Here $n(r)$ is the (differential) distribution function over $(-\log r)$, where r = radius of the particle (approximately considered as spherical), N = total number of particles per unit volume with radius $> r$ and A = a constant. The first equality is the definition of $n(r)$ and the second one (inverse proportionality with r^3) is Junge's law. This law seems to be quite generally valid for continental aerosol. It may be noticed that, as the number of particles is proportional to r^{-3} but the mass of each particle is proportional to r^3, the *mass* distribution is approximately uniform over the range of Junge's law.

It is customary to use the following nomenclature when speaking of aerosol particles:

Name	Size range (μm)
Aitken nuclei or condensation nuclei (CN)	0.005–0.1
Large nuclei	0.1–1
Giant nuclei	> 1

Stratospheric aerosol. A definite layer of large nuclei exists in the stratosphere, at around 20 km altitude. Its origin is still not well understood, but these are particles probably formed by chemical reactions at that level.

Sinks. The efficient processes of elimination from the atmosphere depend on the size range. The largest particles have an appreciable falling velocity and therefore sedimentation is important for them. Also *impaction* (e.g. on leaves of trees) is important for the large particles. Also the particles in the large size range act as nuclei for the condensation of cloud droplets (CCN = cloud condensation nuclei; their role will be studied in Ch.V); when precipitation develops, these particles are carried down to the ground in the raindrops or ice particles (rainout). Large particles can also be scavenged by raindrops (washout). On the other hand, the small particles ($< 0.1\,\mu$) cannot be eliminated in these ways. But they can attach to drops by Brownian motion or other effects (diffusiophoresis, thermophoresis)*: they will also be included in the rainout. Coagulation eliminates smaller particles with production of bigger ones. Figure 10 summarizes these elimination

---

* When there is a concentration gradient in the direction x: $\partial c/\partial x$, for instance of water vapour, the particles in suspension experience a force in the direction of $-\partial c/\partial x$ (i.e. the diffusing water molecules tend to 'drag' along the particles). This effect is called diffusiophoresis. Thermophoresis is a similar effect due to temperature gradients; the force is then in the direction $-\partial T/\partial x$. Both effects act (in opposite sense) during the growth and evaporation of water droplets.

processes. It may be seen that the sizes around 0.1 μm, too small for acting as CCN or to be scavenged by impact, and too big to be captured by Brownian motion, are removed with more difficulty. The figures also indicate the size range over which most of the mass of atmospheric aerosol is distributed, both for the case of continental and of maritime aerosol.

Fig. II-10. *Aerosol elimination processes.* The predominant processes of elimination according to the size range are indicated schematically. The horizontal logarithmic scale indicates the radius of the particles. CCN: activity as cloud condensation nuclei. The ranges of continental and maritime aerosol are also indicated.

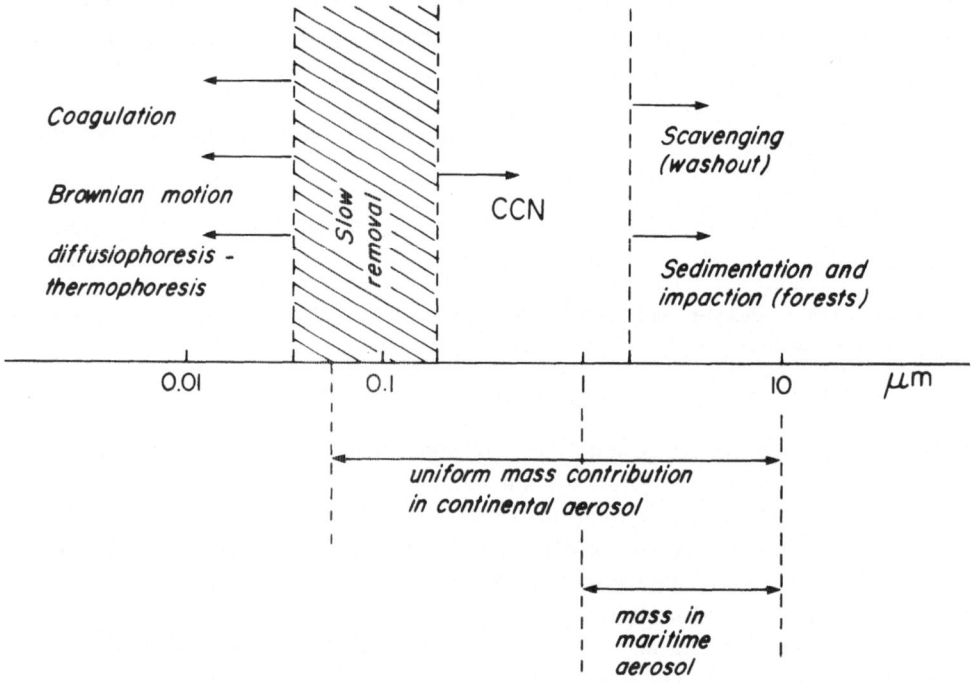

11. Atmospheres of Other Planets

It may be interesting to compare the chemical composition of our atmosphere with that of other planets. Table 5 gives the compositions for the inner and for the outer planets, according to our present knowledge. A remark should be made here. While for the inner planets the atmosphere is clearly differentiated from the solid core of the planet, this is not so for the outer planets. The latter are composed almost entirely of hydrogen and helium, and the transition from the outer parts of the atmosphere to the core is likely to be gradual.

TABLE 5

A. Composition of the atmospheres of the outer planets[1]

(The number of molecules above the clouds in a vertical column with a cross sectional area of one square centimeter)

	H_2	He	CH_4	NH_3
Jupiter	1.8×10^{26}	$< 9.1 \times 10^{25}$	1.2×10^{23}	2.6×10^{22}
Saturn	3.7×10^{26}	$-$[†]	9.4×10^{23}	$< 6.7 \times 10^{21}$
Uranus	1.3×10^{27}	$-$	9.4×10^{24}	$-$
Neptune	$-$	$-$	1.6×10^{25}	$-$

[†] Dashes indicate that no measurement is available, not that the particular gas is absent.

B. Composition of the atmospheres of the inner planets

	Venus[2]	Mars[3]
CO_2	96.4%	95.3%
N_2	3.4%	2.7%
H_2O	0.135%	0.03% (variable)
O_2	69.3 ppm	0.13%
Ar	18.6 ppm	1.6%
Ne	approx. 4 ppm	$-$
SO_2	approx. 186 ppm	$-$
CO	< 0.6 ppm	0.07%

For the Earth, see Table 1.

(1) From Goody and Walker. See Bibliography.
(2) Data obtained by the Pioneer Venus sounder probe (Dec. 9, 1978). From V. I. Oyama, G. C. Carle, F. Woeller and J. B. Pollack, *Science* **203**, 802, 1979.
(3) From T. Owen, K. Biemann, D. R. Rushneck, J. E. Biller, D. W. Howarth and A. L. Lafleur, *J. Geoph. Res.* **82**, 4635, 1977.

Chapter II: Questions

Q1. Different gases are more or less homogeneously distributed in the atmosphere. How do you expect the degree of homogeneity to be affected by the following factors: total abundance, rate of turnover, reactivity, residence time, origin?

Q2. Do you expect Ne to be uniformly distributed? And SO_2? Explain.

Q3. Of all the gases listed in Table 1, which are the ones more easily lost to interplanetary space? Explain.

Q4. Which are the gases affected by photosynthesis in plants?

Q5. Why is the lower stratosphere so dry?

Q6. Could the very dry air of the lower stratosphere in middle latitudes have come from the troposphere after crossing the tropopause in the polar regions?

Q7. Do you expect that the presence of SO_2 will increase or decrease the pH of cloud water? The presence of NH_3? Explain.

Q8. Among the nitrogen oxides, NO oxidizes readily in the atmosphere to NO_2. Does N_2O undergo a similar oxidation?

Q9. Will the presence of CO_2 help the dissolution of NH_3 in cloud droplets? Explain.

Q10. Why is solar radiation important in the formation of 'smog'?

Q11. The water cycle in the atmosphere plays an important role in the evolution of some trace gases. Which are these gases?

Q12. Is the water cycle in the atmosphere linked to that of aerosol particles?

Q13. Why are aerosol particles in the range around 0.1 μm removed with more difficulty than smaller and larger particles?

Chapter II: Problems

(Any necessary constants not given in the statement of a problem will be found in the Table of Constants on pages x–xi)

P1. Knowing that the average concentration of neon in the atmosphere is 18 ppm (in volume), calculate the total mass of neon in the atmosphere.

P2. Express the composition of dry air in mass percentages, for the four main constituents. Use the data of Table 1 and assume ideal gas behaviour.

P3. Concentrations of trace gases are usually expressed either in ppm (or ppb) by volume or in $\mu g/m^3$. Find the equivalence, for sulfur dioxide SO_2, at 1 atm pressure and $0°C$.

P4. Assume that the mass mixing ratio of water vapour r (mass of water vapour per unit mass of air) has the values (a) 2 g/kg between 1000 mb and 900 mb; (b) 0.2 g/kg between 300 and 200 mb; and (c) 0.002 g/kg above the tropopause, situated at 200 mb. What are the values of 'precipitable water' (thickness of liquid water that would be obtained if all the water vapour were condensed) in these three regions, in mm? The third region may be considered as extending indefinitely in height, with constant r.

P5. If air at 1 atmosphere is saturated with water vapour at $-10°C$, calculate:
   (a) its mass mixing ratio, in g/kg;
   (b) its volume mixing ratio, in ppm.
   Use Figure 1 to find the saturation vapour pressure.

P6. Consider 1 kg of air. What mass of water vapour, in grams, does it contain if it is saturated (a) at the ground, with a temperature of $25°C$ and a pressure of 1000 mb; (b) at the tropopause in middle latitudes, with a temperature of $-50°C$ and a pressure of 200 mb; and (c) at the tropical tropopause, with a temperature of $-80°C$ and a pressure of 90 mb? The vapour pressure of water at $25°C$ is 31.67 mb; the vapour pressure of ice at $-50°C$ is 3.9×10^{-2} mb, and at $-80°C$ is 5.5×10^{-4} mb.

P7. Knowing that the saturation vapour pressure at $20°C$ is 23.37 mb and that the latent heat of vaporization is 2.5×10^6 J/kg, calculate the saturation vapour pressure at $0°C$.

P8. Assuming that the total input F of hydrogen sulfide into the atmosphere per year is about 30 Tg, expressed as H_2S (cf. Figure 5) and assuming a roughly estimated average concentration of 0.2 ppb, make an estimate of the residence time of H_2S before being converted into other compounds or eliminated from the atmosphere.

P9. If hydrogen sulfide is oxidized and, through a series of reactions, transforms finally into ammonium sulfate, what mass of ammonium sulfate is produced from each kilogram of hydrogen sulfide?

P10. Using the average inflow rate F of ammonia into the atmosphere given in Figure 6 and assuming that the global average concentration in the atmosphere (counting both land and oceans) is 1 ppb, obtain an estimate of the residence time τ.

P11. If a continental aerosol contains 10^4 particles of radii $\geq 0.1 \, \mu m$ per cm^3, how many particles would you expect it to have, with radii $\geq 1 \, \mu m$?

45

III. Radiation

(a) Laws

1. Regions of Electromagnetic Spectrum. Absorption of Radiant Energy

Figure 1 represents schematically the different regions of the spectrum of electromagnetic waves, corresponding to the different regions of wavelength λ and frequency ν. These two quantities are related by

$$c = \nu\lambda \tag{1}$$

where c = velocity of light in vacuum = 2.998×10^8 m/s. Electromagnetic waves carry energy, which may be absorbed by matter, producing different effects. This absorption can occur only in multiples of a *quantum* of energy, proportional to the frequency of the radiation; thus for radiation of frequency ν, the value of the quantum is:

$$h\nu = hc/\lambda \tag{2}$$

where $h = 6.62 \times 10^{-34}$ J.s is the Planck constant. A quantum of radiant energy is also called a *photon*.

In order to understand the effects that absorption of radiant energy (such as carried by the sunlight) may have on the molecules (such as those of the gases that make up atmospheric air), we must still recall some other facts. Molecules can exist only in certain states of rotation, vibration and electronic configuration, with characteristic energies. In other words, the rotational, vibrational and electronic energies of molecules are quantized. The differences between different levels of rotation are much smaller than the differences between different levels of vibration, and these are smaller than those between electronic states. The absorption of electromagnetic radiation occurs by elementary processes, each one involving the absorption of one photon and a simultaneous jump of the molecule from its initial state to another state of higher energy. In the absence of any such possible transition, radiation is not absorbed (although it may be scattered). Thus a quantum of low energy can produce an increase in rotational energy; photons of higher energy will be required to increase the vibrational energy and photons of still higher energy will be required to produce changes in the electronic state. Of course, the last changes can be accompanied by simultaneous changes in rotation and vibration, which require less energy. If the photon is energetic enough, its absorption can result in the dissociation of a molecule or in the ionization of an atom or a molecule. In any case, between the energy ΔE absorbed (i.e. the change of energy of the absorbing system) and the frequency of absorbed radiation, the following relation must hold:

$$\Delta E = h\nu \tag{3}$$

For all the processes without ionization or dissociation, only certain frequencies can be absorbed, corresponding to the values ΔE between different states of the molecule: the absorption is discrete, and can be revealed through absorption spectra showing discrete lines at certain frequencies. On the other hand, translational energy is not quantized. Therefore, when ionization or dissociation occurs, the products (ion and electron, or two atoms or fragments of a molecule) can have any value of translational energy. It results that these processes require a threshold of frequency (a minimum value of ΔE necessary for the ionization or dissociation, according to (3)), but photons of any energy above that threshold can be absorbed, the excess energy remaining as kinetic energy of the products. This results in an absorption spectrum showing continuous absorption above a certain frequency minimum*.

Figure 1 shows schematically, at the bottom, the type of processes that occur by absorption of electromagnetic radiation of the wavelengths and frequencies indicated above.

If the energy of the photon, absorbed by one molecule, is multiplied by Avogadro's number ($N_A = 6.0 \times 10^{23}$), we have the energy referred to one mole, rather than one molecule. These values are also indicated in Figure 1 in Joules per mole.

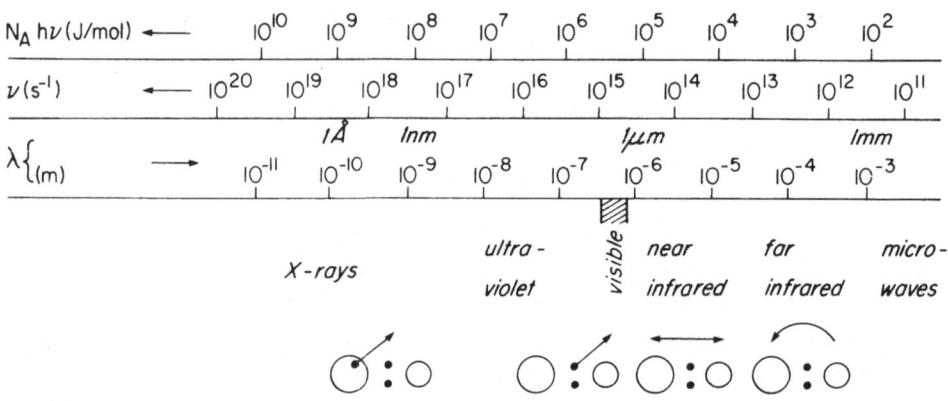

Fig. III-1. *Electromagnetic spectrum*. The scales indicate: the energy of one 'einstein' or mole of photons $N_A h\nu$ (N_A = Avogadro's number; h = Planck's constant), the frequency ν and the wavelength λ. The spectrum extends farther to the left of the figure (γ-rays) and farther to the right (radio waves). The sketches indicate the type of action a photon of that range can have on a molecule; from right to left (increasing energy): increase of rotational energy; of vibrational energy; electronic excitation, dissociation or ionization, involving valency electrons; electronic excitation or ionization, involving inner electrons.

2. Black Body Radiation. 'Short Wave' and 'Long Wave' Radiation in the Atmosphere

We have seen how the absorption of radiant energy is associated with transitions of molecules from a certain state to another state of higher energy. Similarly, the transition to a state of lower energy can be associated with the emission of a photon of the corresponding frequency. Thus any body can emit as well as absorb radiant energy. We shall say that a body is in radiative equilibrium with its environment if it emits as many photons as it

---

* Notice that no absorption occurs for which the primary effect would be a simple increase in the kinetic energy of a molecule.

48

absorbs, per unit time, for each particular interval of frequencies. We shall call *radiant emittance* or *emissive power* E the total radiant energy emitted per unit surface and unit time. The radiant emittance at a given wavelength E_λ will be the radiant energy emitted per unit surface, unit time, and unit interval of wavelength, in the interval between λ and $\lambda + d\lambda$ (or the corresponding interval of frequencies)*. Thus

$$E = \int_0^\infty E_\lambda \, d\lambda \tag{4}$$

For reasons that will become apparent later, we are particularly interested in the radiation emitted by a particular kind of bodies: the so-called 'black bodies'. Their radiation is called 'black body radiation'. But before describing the laws that govern this type of radiation, let us give some definitions. We define as *isotropic* any radiation that is equal in composition and intensity for any direction in space. If isotropic radiation of wavelength λ is falling upon the surface of a body, part of it can be absorbed, part reflected and part transmitted. We shall call *absorptivity* A_λ the fraction that is absorbed by the body. We can equally define an integral absorptivity A for radiation integrated over the whole spectrum of wavelengths. The absorptivity A_λ depends on the nature of the body and on its temperature.

Bodies with absorptivities close to unity for every λ can be prepared. Better than that, surfaces equivalent to a surface of a body of absorptivity practically equal to unity can be obtained in the form of an opening in a hollow, internally blackened cavity. Any radiation coming into the cavity through the opening becomes virtually trapped, because it is reflected many times in the cavity before it can find its way out through the opening, and a fraction close to unity is absorbed at each one of the reflections. We define a *black body* as one which has unit absorptivity at any wavelength. Therefore, the opening is equivalent to an equal area of a black body. Inside the cavity, a radiative equilibrium is established, whereby atoms or molecules of the surface emit as many photons of each wavelength per unit time as they absorb. A small part of the emitted radiation finds its way out through the opening; this will be the characteristic radiation of a black body. It will be described by its radiant emittance E_{λ_b} as a function of λ (the subscript b indicating that it corresponds to a black body). The black body radiation is independent of the nature of the body; this fact suggests to us its fundamental importance. The laws of black body radiation have been studied experimentally and theoretically and have had great importance in the development of physics. Here we shall only state them.

(1) Black body radiation is isotropic and depends only on the temperature of the body (but not on its nature).

(2) The integral radiant emittance E_b is proportional to the fourth power of the temperature (Stefan–Boltzmann law):

$$E_b = \sigma T^4 \tag{5}$$

where $\sigma = 56.7 \, \text{nW/m}^2 \, \text{K}^4 = 1.33 \times 10^{-12} \, \text{cal/cm}^2.\text{s}.\text{K}^4 \, (1 \, \text{nW} = 10^{-9} \, \text{W})$.

---

* This is not to be confused with the intensity I_λ or brightness, which is referred to the unit of solid angle. E_λ and E refer to the total emission in any direction, over a hemisphere above the unit area. Between emittance and intensity, and isotropic radiation, exists the relation $E_\lambda = \pi I_\lambda$.

(3) The radiant emittance at a given λ, defined by $E_{\lambda_b} = dE_b/d\lambda$, is given by Planck's law:

$$E_{\lambda_b} = \frac{2\pi c^2 h}{\lambda^5} \frac{1}{e^{hc/\lambda kT} - 1} \tag{6}$$

where $k = 1.38 \times 10^{-23}$ J/K is Boltzmann's constant and the other symbols have the same meaning as before.

The Stefan–Boltzmann law (5) can be considered as a consequence of Planck's law (6), from which it can be obtained directly by integration. Planck's law gives the distribution of the energy emitted by a black body over the wavelength. It gives curves of the type of Figure 2, where the three curves correspond to three different temperatures. Several

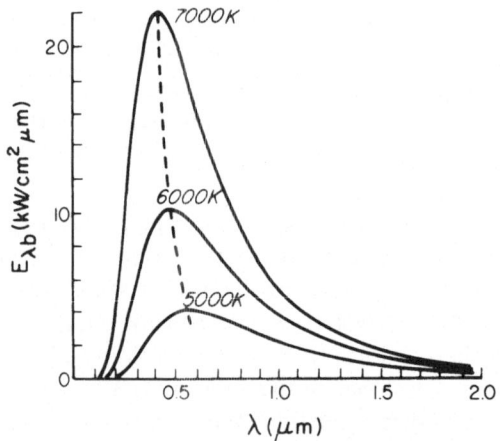

Fig. III-2. *Black body radiation.* E_{λ_b} = radiant emittance at wavelength λ (Planck's formula), for three temperatures. The dashed line joins the three maxima, to illustrate Wien's displacement law.

remarks can be made about this figure. It can be observed that the maximum becomes shifted towards the short wavelengths as the temperature rises; in fact, the values for λ at the maximum varies according to the law

$$\lambda_m . T = \text{const} = 0.288 \, \text{cm}.\text{K} \tag{7}$$

Also, the value E_{λ_m} for the maximum can be shown to be proportional to T^5. These two results are sometimes referred to as Wien's 'displacement laws'. They give the reason why the glow of incandescent bodies starts with a dark red colour and becomes both whiter and brighter as their temperature rises.

(4) We add now a fourth law of radiation, called Kirchhoff's law: the radiant emittance of any body (non-black in general) is equal to its absorptivity times the radiant emittance of the black body, at the same temperature:

$$E_\lambda = A_\lambda . E_{\lambda_b} \tag{8}$$

In other words, the radiant emittances are proportional to the absorptivities, the absorptivity of the black body in particular being equal to unity:

$$\frac{E_{\lambda_1}}{A_{\lambda_1}} = \frac{E_{\lambda_2}}{A_{\lambda_2}} = \ldots = \frac{E_{\lambda_b}}{1} \qquad (9)$$

where the subscripts 1, 2, etc. indicate different bodies (all at the same temperature). Thus bodies that absorb more in a certain wavelength also emit more in the same wavelength.

The ratio $\epsilon_\lambda = E_\lambda/E_{\lambda_b}$ is called the *emissivity*. Thus the emissivity of a black body is equal to unity, and in general, for any type of body

$$\epsilon_\lambda = A_\lambda \qquad (10)$$

i.e. the emissivity is equal to the absorptivity.

The interest that black body radiation has for us lies in that: (1) the Sun emits approximately as a black body at $\sim 6000\,\text{K}$, at least in the main part of the solar radiation spectrum; (2) the Earth, and also clouds, behave as black bodies, at their own temperatures.

Due to the large difference between the temperature of the Sun and, say, the average temperature of the Earth's surface, the two spectra of radiant emittances are displaced from each other far enough to have practically no superposition. Also, the total emittance, and therefore the areas below the curves, will be in the ratio of approximately $(6000/273)^4 \cong 2 \times 10^5$ (see Equation (5)). However, only a very small solid angle of the radiation emitted by the Sun is covered by the Earth, and the energy absorbed must be exactly compensated, in the yearly average, by the energy emitted by the Earth. Therefore, if we draw the two spectra referred to the total average incoming and outgoing radiation, rather than to unit areas, we have a representation such as given in Figure 3. Practically the whole solar spectrum lies at $\lambda < 4\,\mu\text{m}$ and the whole terrestrial spectrum at $\lambda > 4\,\mu\text{m}$, so that $4\,\mu\text{m}$ can be taken as the limit separating both regions. Thus *solar radiation* or *'short wave' radiation*, and *terrestrial radiation* or *'long wave' radiation*, can be considered separately.

(b) Solar Radiation

3. Absorption of Solar Radiation in the Atmosphere

Except for weak lines of excitation[*] of O_2 and some absorption by O_3 (Chappuis bands), little absorption occurs in the visible portion of the solar spectrum. Absorption becomes very important however in the ultraviolet region ($< 370\,\text{nm}$), so that very little of this radiation reaches the ground.

In order to describe how this radiation is absorbed and what are the effects, let us consider the incoming solar radiation, penetrating the atmosphere of the Earth. The more energetic radiation (shorter wavelengths) will be quickly absorbed by atoms or molecules that become ionized or dissociated. Less energetic radiation will be able to penetrate deeper, until it finds enough concentration of gaseous species capable of absorbing it. In general, the more it penetrates, the higher the density of the absorber (because the press-

---

[*] By 'excitation' is understood the transition of the molecule (or the atom, as the case may be) from a lower to a higher state of energy.

ure increases rapidly towards the ground), so that once radiation of a certain wavelength starts being absorbed by some of the air components, it becomes quickly exhausted. However, it must be remembered that the ozone, which is a very important absorber, is

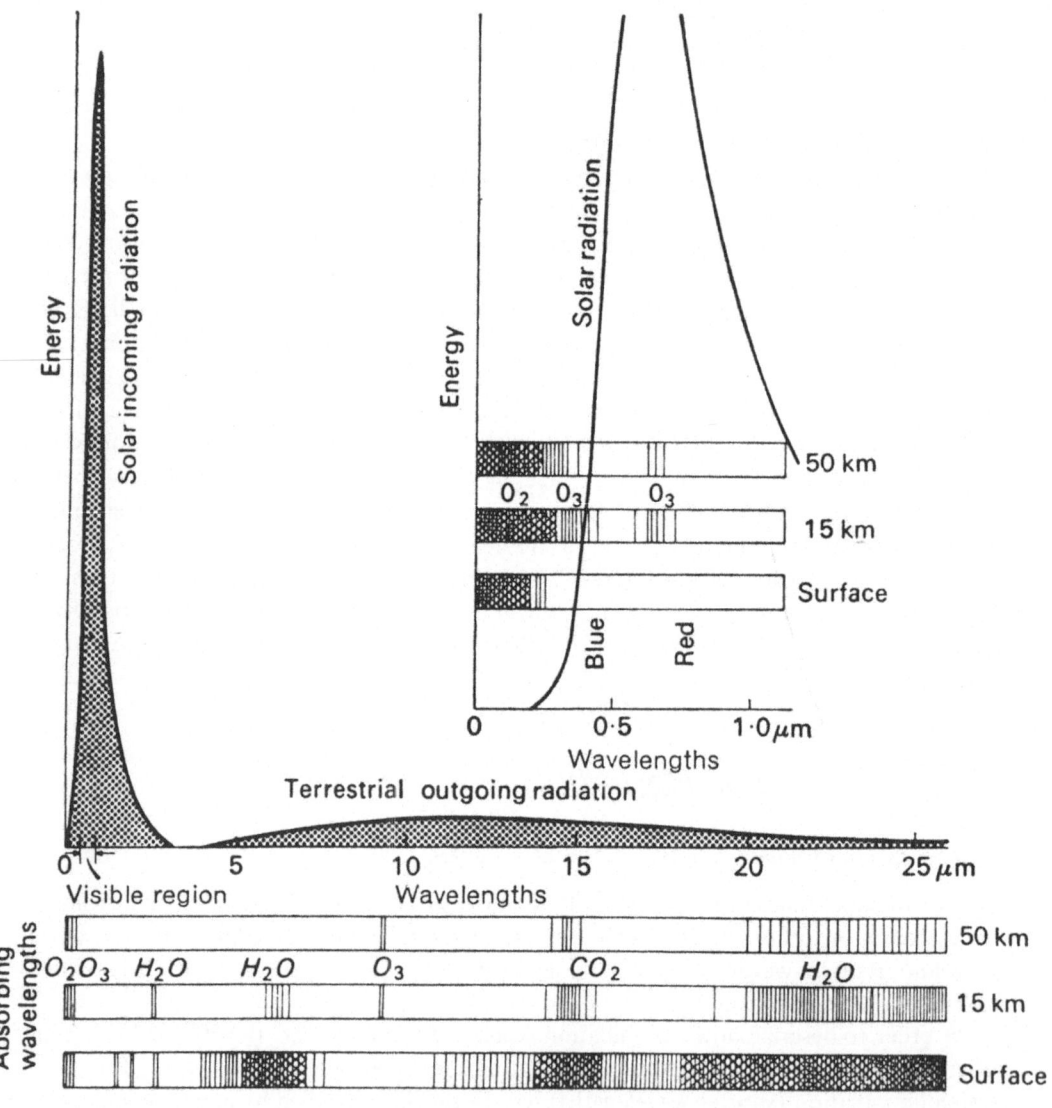

Fig. III-3. *Spectral distribution of solar and terrestrial radiation.* The curves marked solar and terrestrial radiation indicate the distribution of energy in the different wavelengths of radiation from these two sources. Note that the energy of the incoming solar, and the outgoing terrestrial radiation lie in quite separate parts of the spectrum. (Blue light has a wavelength of about 0.4 μm and red light, about 0.7 μm.) The shaded strips below the curves indicate the wavelengths where the atmosphere absorbs radiation. Note that there is much absorption of the outgoing terrestrial radiation by the lower layers but little by the highest layers. The small inset shows the short-wavelength end of the diagram with a more open wavelength scale.

concentrated at a certain level, i.e. its concentration passes through a maximum instead of increasing continuously for decreasing altitude. The situation for most absorbers will be such as is illustrated schematically in Figure 4, where the three curves represent the inten-

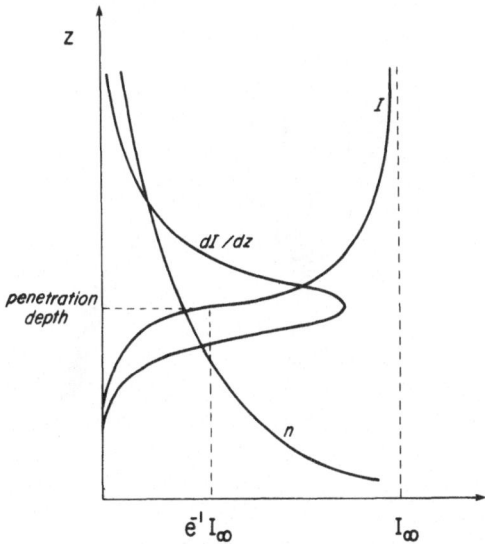

Fig. III-4. *Absorption of short-wave radiation in the atmosphere.* The graph shows schematically the relative magnitudes, as a function of height z, of abundance of absorber (n = number concentration of molecules), rate of absorption dI/dz at each level and intensity I of the remaining radiation as it travels downward; I_∞ is the value of incident radiation at the outer edge of the atmosphere. See text in this section for general explanation and §4 for detailed treatment.

sity (flux density) of radiation of a given wavelength, the rate at which it is being depleted on its way downwards, and the concentration of the absorber. The rate of depletion gives also the rate at which the products of the photochemical reaction appear at each level. The atmospheric absorption at a given wavelength is characterized by the *penetration depth*, which is defined as the altitude at which, for vertical incidence, the intensity of the radiation has been reduced by a factor $e^{-1} = 0.37$. For the main atmospheric constituents, whose abundances decrease exponentially with height, this altitude also corresponds to the level of maximum rate of absorption. This will be demonstrated in §4. Figure 5 shows the penetration depth as a function of the wavelength, and the main absorption processes are indicated on it.

We can describe briefly these processes as follows:

– Close to 300 nm (3000 Å) O_3 starts absorbing (Huggins band) (and thereby dissociating into O_2 + O: see §6), and continues doing so for the interval between 200 and 300 nm (Hartley band). The result of this absorption is a virtual cutoff in the wavelengths of radiation arriving to the ground at about 300 nm. The penetration depth for λ between 200 and 300 nm is about 40 km (see Figure 5).

– Below 200 nm O_2 absorbs effectively (Herzberg and Schumann–Runge regions) and dissociation occurs. O atoms thus appear and become abundant above 80 km.

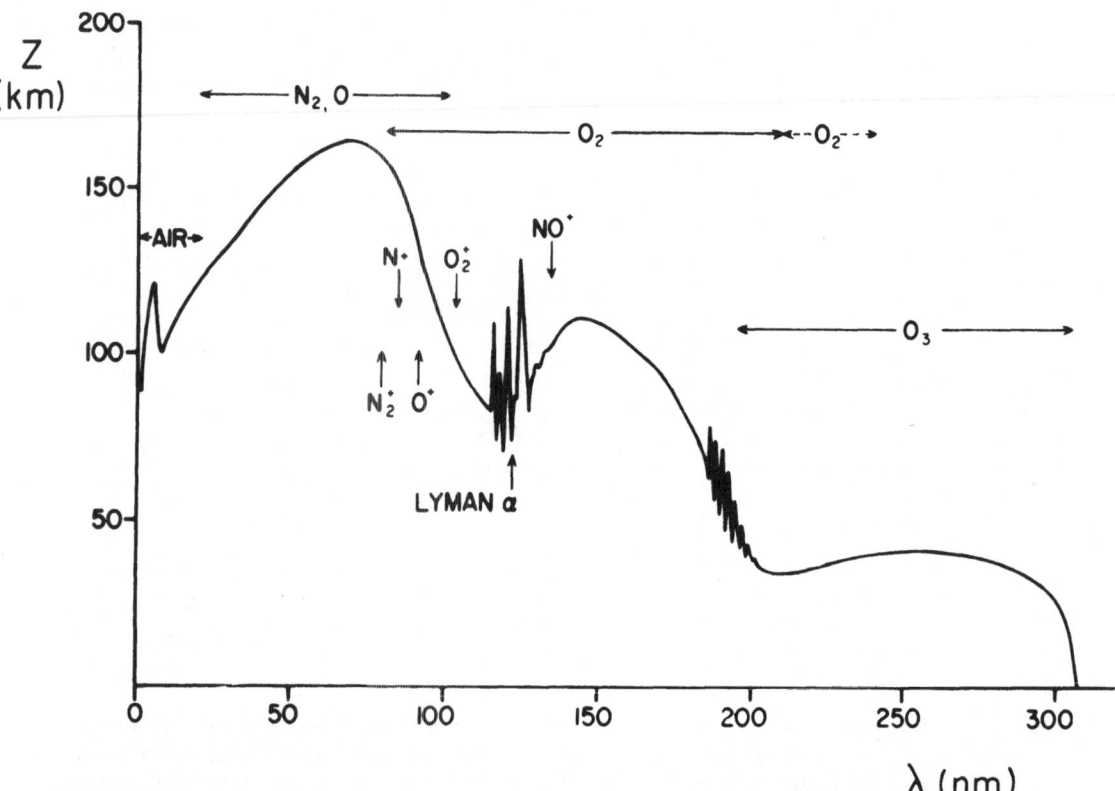

Fig. III-5. *Absorption of solar radiation by the atmosphere.* z = penetration depth, i.e. altitude at which the intensity of solar radiation drops to e^{-1} of its value outside the Earth's atmosphere, for vertical incidence. λ = wavelength. The horizontal arrows at the top indicate the ranges of absorption for the corresponding molecules. The ions with vertical arrows indicate the ionization thresholds. (Copyright © 1965 by Prentice-Hall, Inc.)

— Below 150 nm, the photons become successively (as λ decreases) energetic enough to be able to ionize NO, O_2, O, N, N_2:

$$NO + h\nu \rightarrow NO^+ + e \quad (\lambda < 134.1 \text{ nm})$$
$$O_2 + h\nu \rightarrow O_2^+ + e \quad (\lambda < 102.6 \text{ nm})$$
$$O + h\nu \rightarrow O^+ + e \quad (\lambda < 91.0 \text{ nm})$$
$$N + h\nu \rightarrow N^+ + e \quad (\lambda < 85.2 \text{ nm})$$
$$N_2 + h\nu \rightarrow N_2^+ + e \quad (\lambda < 79.6 \text{ nm})$$

(11)

The corresponding thresholds are indicated in the graph.

— The penetration depth in the region 100–150 nm is seen to vary rapidly with λ, indicating certain 'windows' or intervals of wavelength for which radiation is less absorbed and therefore is able to penetrate down to about 70 km.

— N_2 dissociates only by the action of radiation with $\lambda < 127$ nm. But this is not a direct dissociation; it occurs only by a different process called 'predissociation', which consists in the activation of the N_2 molecule to an unstable state, followed by dissociation:

$$N_2 + h\nu \rightarrow N_2^*$$

$$N_2^* \rightarrow 2N \qquad\qquad (12)$$

All these photochemical actions of the solar radiation alter the composition of the upper layers of the atmosphere. They combine with the diffusion process studied in Ch.I to produce the variable composition of the heterosphere (as opposed to the uniform composition of the homosphere, below 80–100 km). Figure 6 shows a model of composition of the thermosphere (a) and exosphere (b) estimated for a temperature of about 1400 K

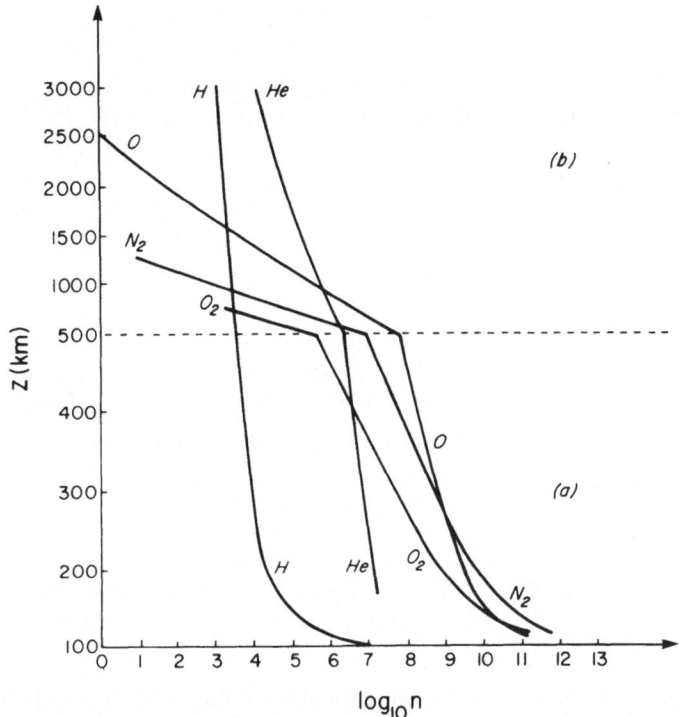

Fig. III-6. *Distribution of the main neutral constituents in the thermosphere (a) and exosphere (b).* Notice the change of vertical scale at 500 km.

in the higher constant temperature region (cf. Figure I, 4); only the main constituents are represented. Several remarks can be made here.

As the concentration of constituent gases is given in a logarithmic scale of their number density of molecules, a constant ratio between two gases along height (as exists in the homosphere) would be given by two parallel lines (constant difference in the logarithm). Thus the divergence of lines indicates the variations in relative abundances due to both photochemistry and diffusion. The tendency of the lines to become straight in the higher levels indicates the predominance of diffusion as the controlling factor in vertical distribution; the slope at any level is then given by $1/H$. The scale height H only varies slowly in this approximately isothermal region (cf. Ch.I, §2).

In particular, we can observe that the relative O concentration increases rapidly above 100 km, becomes more abundant than O_2 above about 130 km and the predominant species above 250 km. This is due to the dissociation of O_2 (Schumann–Runge region of wavelengths, cf. Figure 5) in the region of 100 km and diffusion above that level.

In the exosphere diffusion continues sorting out the lighter species with increasing height, so that both H and He become predominant over O above 1500 km.

It should also be remarked that all the photochemical processes which we have been describing are responsible for the high temperatures above 100 km (thermosphere). This is apparent through the diurnal cycle: much higher temperatures are attained during the day (when solar radiation is being absorbed) than during the night. Thus above 300 to 500 km, the temperature becomes approximately constant with height, but this constant temperature can vary in the range of 500–1500 K during the night, and in the range of 1000–2000 K during the day, depending on the activity of those processes.

Here we have restricted our consideration to the neutral species. Photochemistry is also responsible for the production of ionization, and this will be studied in §5.

4. The Chapman Profile*

Although the absorption of ultraviolet radiation through the high atmosphere is a complicated problem, the essential features of the process can be obtained from a simple treatment under simplified assumptions. This was first done by S. Chapman, and we shall call the curve giving the rate of absorption of radiation as a function of height a Chapman profile.

We shall restrict the derivation to the simple idealized case defined by the following assumptions:

a) The radiation is monochromatic (i.e. of a given λ) and its radiant flux density incident on the atmosphere from outside (energy per unit time and unit cross section, coming from the Sun) is I_∞.

b) There is only one species of molecule capable of absorbing that radiation, bearing an exponential distribution in height (such as would be expected for an isothermal region in diffusive equilibrium):

$$n = n_0 \, e^{-z/H} \tag{13}$$

Here n is the number density of molecules, n_0 its value at $z = 0$, which defines a certain reference level from which the height z is measured, and H is a constant scale height (see Ch.I, §2).

c) The radiation falls vertically onto the atmosphere, i.e. the zenith angle is 0.

___

* This section can be omitted without loss of continuity.

d) *Beer's law* for the absorption of radiation is applicable. According to this law, the decrease dI in radiant flux density I after travelling a thickness dz is proportional to I and to dz:

$$dI = I\alpha n \, dz \tag{14}$$

where α is a constant called the *absorption cross section* of the absorber molecules. If we consider the radiation travelling down, both dI and dz are negative.

The rate at which the radiation is being absorbed at a given height, which is also the rate at which the products of the photochemical process will appear, is given by

$$q = \frac{dI}{dz} = I\alpha n \tag{15}$$

Let us consider the total number of absorber molecules existing above a level z; this will be given by (using Equation (13)):

$$n_t = \int_z^\infty n \, dz = n_0 H e^{-z/H} \tag{16}$$

Integration of (14) gives

$$I = I_\infty \exp\left(-\alpha n_0 H e^{-z/H}\right) = I_\infty e^{-\alpha n_t} \tag{17}$$

Introducing (13) and (17) into (15), we derive the profile $q = q(z)$ as

with
$$q = q_0 \exp\left[-\left(\frac{z}{H} + \alpha n_0 H e^{-z/H}\right)\right], \tag{18}$$

$$q_0 = \alpha n_0 I_\infty$$

This function becomes a maximum when the exponent has a minimum absolute value. By differentiating it and equating to 0, this is found to occur when

$$z = z_m = H \ln (\alpha n_0 H) \tag{19}$$

which, when introduced in (17), gives

$$I = I_\infty e^{-1} \tag{20}$$

Equation (20) demonstrates for the assumed model the previous statement that the penetration depth (defined by condition (20)) is also the level at which the rate q of radiation absorption is a maximum.

Expression (18) gives the desired Chapman profile and it is represented in Figure 7. The altitude is expressed as a normalized height $(z - z_m)/H$, where z_m is given by (19), and the absorption rate is normalized against the maximum absorption rate $q_m = e^{-1}I_\infty/H$ (value obtained by introducing (19) into (18)).

As mentioned before, the rate of radiation absorption q is also a measure of the rate of production of the products of the photochemical reaction. If this is ionization of a molecule or an atom in the ionosphere, it will give the rate of production of free electrons. In order to obtain this figure in terms of number of particles (electrons, ions, or atoms as the case may be) produced per unit time and unit volume, q should be divided by the energy $h\nu$ of a photon.

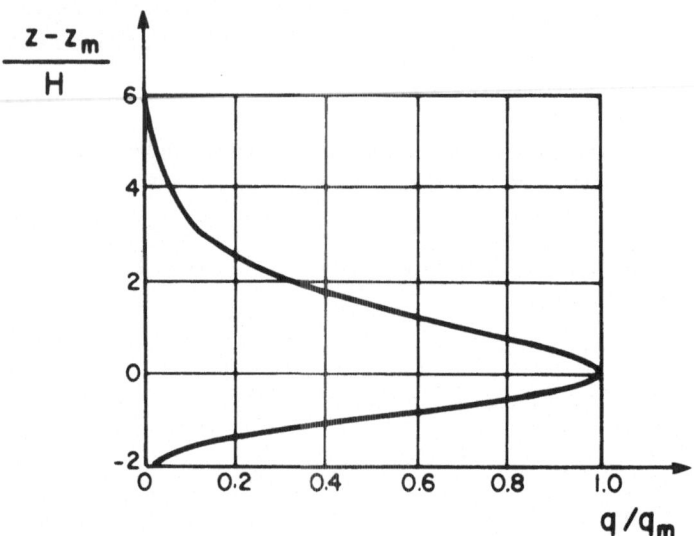

Fig. III-7. *Chapman profile*. It gives the rate of absorption of monochromatic radiation q with height z in an atmosphere where the concentration of absorber decreases exponentially with height. The curve is plotted in terms of normalized height $(z - z_m)/H$ (where z_m = height for maximum q; H = scale height) and rate of absorption q/q_m (q_m = maximum value of q).

5. Photochemistry of the Ionosphere

We described briefly in Ch.I, §4 the main characteristics of the ionosphere, in particular the different regions identified in it. We are now better equipped, with the notions already summarized in the present chapter, to explain the main points regarding the interpretation of these characteristics.

As can be followed on Figure 5, the shorter wavelength radiation, more energetic and therefore able to ionize all molecular and atomic constituents, is absorbed in the higher levels, so that only radiation of longer wave or that corresponding to some special intervals ('windows') is able to penetrate to lower altitudes. We shall later consider the situation separately for the different regions, but first we shall describe some general characteristics of the photochemistry of the ionosphere.

The primary processes consist in the ionization of the neutral species present by radiation of sufficient energy. The main ionizations were already given in Equations (11). These primary ions may then undergo different sorts of reactions with the neutral molecules. Examples are:

$$O^+ + O_2 = O + O_2^+ \qquad \text{(charge transfer)}$$

$$O^+ + N_2 = NO^+ + N \tag{21}$$

Ions finally disappear, and the description of the conditions in a region must consider this aspect as well. Chemically, ions will disappear by recombining with free electrons. However, atomic ions can do that only in three-body collisions:

$$O^+ + e + M = O + M \tag{22}$$

where M stands for any kind of molecule or atom ('third body'). M does not react; its role is to carry away, as kinetic energy, the excess energy released by the recombination of ion and electron, in such a way that both energy and momentum are conserved. Otherwise, the recombined atom remains with enough energy to redissociate immediately again. The frequency of such triple collisions is, of course, much smaller than that of two-body collisions, and will vary rapidly with the number concentrations of the particles involved; in fact the *law of mass action* establishes that the rate at which the previous reaction occurs (as measured, for instance, by the number of O^+ which disappeared or the number of O formed per unit time and unit volume) must be proportional to the concentrations of O^+, e and M. As the concentration of neutral M decreases very rapidly with height, we must expect that recombination of atomic ions becomes first a rare event and then ceases to occur.

The situation is different for molecular ions. With them, the energy and momentum surplus in a two-body recombination may be used in dissociating the molecule. Thus, this type of 'dissociative recombination' constitutes an important sink of ions in the ionosphere. Examples of such reactions are:

$$O_2^+ + e = O + O$$

$$NO^+ + e = N + O$$

$$N_2^+ + e = N + N \tag{23}$$

Notice that atomic ions, at levels where (22) cannot occur, may disappear by a combination of reactions such as the first (21) followed by the first (23).

If no other factors were to be considered, the sequence of photochemical ion production, chemical reactions and recombination could lead, if occurring fast enough, to an approximate steady state (i.e. constant concentrations and distribution of the species involved) during periods of approximately constant radiation influx; when this occurs, we speak of conditions of *photochemical equilibrium*. In general, however, the processes of mixing and diffusion must also be considered. The final result depends on the relative speeds of the three types of processes (ion reactions and the two just mentioned). Ion reactions are, in general, fast and even reaction (22) is faster than mixing at the relevant levels; we know also that mixing only occurs below about 100 km. Thus, the two really competing processes are photochemistry and diffusion. Chemical reactions become slower with lower pressure, while diffusion becomes faster; so we may expect that diffusion becomes the controlling factor for ion distribution when the altitude is high enough. We shall see that this is the case above 300 km (F region).

Let us now consider separately the different regions.

D region. This layer, being the lowest, is effectively shielded by the upper layers from most of the solar short wave radiation which is capable of ionizing air molecules. Radiation with λ from 122 to 180 nm has been spent in dissociating the O_2 molecule and, with $\lambda < 102.6$ nm, in ionizing the same molecule (cf. Figure 5). Radiation with $\lambda > 180$ nm is no longer able to produce ionization. However, in the region 102–122 nm there are certain windows, i.e. intervals, in which no absorption occurs above the D region. In particular, a hydrogen line (called Lyman α) of $\lambda = 121.6$ nm reaches these levels through one of the windows and is able to ionize NO and is thus largely responsible for the existence of the D region.

In disturbed conditions, particularly during SID (cf. Ch.I, §4), appreciable amounts of X-rays with $\lambda < 0.6$ nm reach the D region level and produce intense ionization of any sort of air molecules. This is responsible for blackout in radio communications due to the enhanced absorption of waves in this region, as explained in Ch.I, §4. In normal conditions, X-rays also contribute to the ionization of the upper D region, and cosmic rays to that of the lower part.

The formation of negative ions is a characteristic of the D region. They are produced by electron attachment, and such a reaction requires the presence, during the collision, of a third body M for the same reasons explained above with regard to the neutralization of atomic ions (reaction (22)):

$$O_2 + e + M = O_2^- + M \tag{24}$$

Here the concentrations of both O_2 and M will decrease rapidly with increase of height, like the pressure. In consequence, one must expect that negative ions will form predominantly in the lowest layers of the D region, and their production will decrease rapidly with height. They are virtually absent in the other ionospheric regions.

The primary ions formed in this region (mainly positive NO^+ and O_2^+ and negative O_2^-) can undergo subsequent reactions with a large variety of neutral molecules, making of the D region a complex system with many ionic species. For instance, the pressure is high enough to permit the formation of some 'cluster ions', like $NO_3^-(H_2O)_n$, where a number of neutral molecules have become attached to the ion itself. A summary of the main ionic constituents of the different regions can be found in Table 1, which includes approximate values of their concentrations (as the decimal logarithm of their number concentrations of ions). This is also illustrated in Figure 8.

TABLE 1

Composition of the ionosphere in daytime (quiet conditions)

Region	Altitude (km)	Typical temperature (K)	Principal ions (Log$_{10}$ (number concentration in cm^{-3}))
D	60–90	250	NO^+: 3.75 $H_3O^+ \cdot H_2O$: 3 H_3O^+: 2.5 $NO_3^-(H_2O)_n$
E	110	250	NO^+: 4.5 O_2^+: 4.5
F_1	170	700	NO^+: 4.5 O_2^+: 4.5 O^+: 5.5
F_2	300	1500	O^+: 5.5 N^+: 4 O_2^+: 3 NO^+: 3

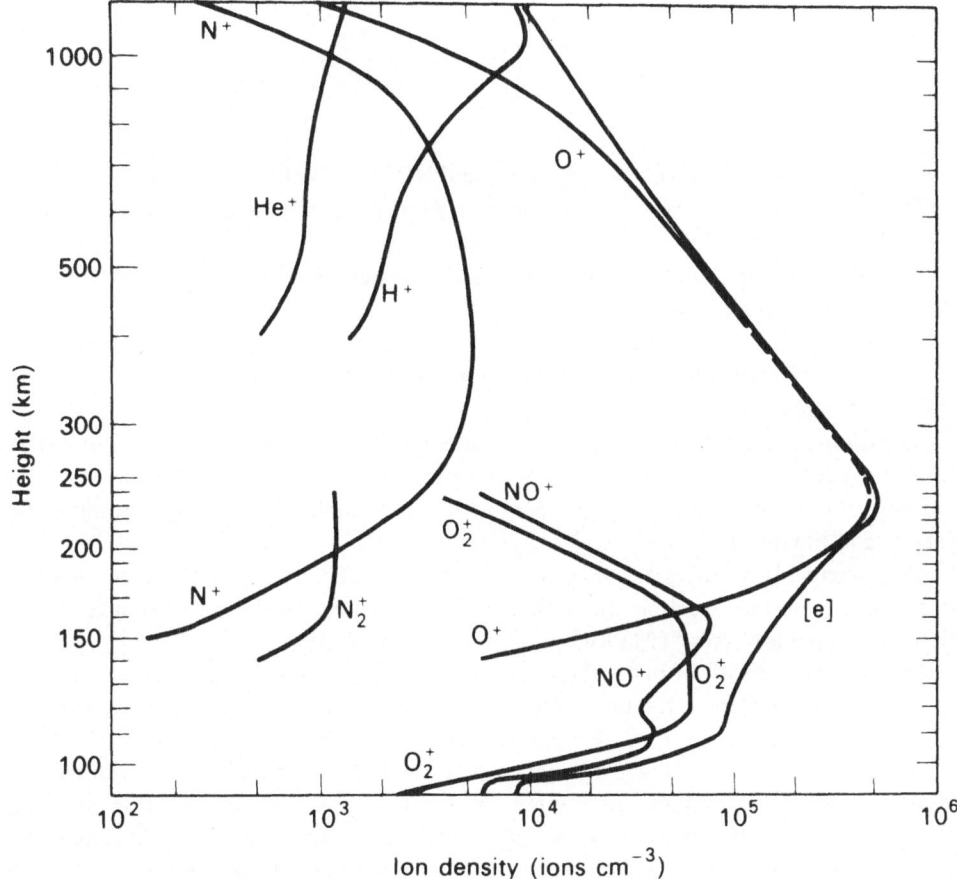

Fig. III-8. *Positive ion composition of a solar minimum, daytime ionosphere.* The ion distributions shown are based on results from two mass spectrometer experiments. The data has been normalized to the electron density distribution (*e*) measured during the same general period.

The loss processes are mainly represented by the dissociative recombinations described by Equation (22). Electrons are also lost by the three-body reactions producing negative ions, as mentioned above.

All these processes are much faster than mixing or diffusion, so that they control the ion composition of the region. The *D* region is therefore essentially in photochemical equilibrium.

E Region. Enough radiation penetrates into this region, capable of ionizing O_2 ($\lambda < 102.6$ nm) and O ($\lambda < 91.1$ nm):

$$O_2 + h\nu = O_2^+ + e$$

$$O + h\nu = O^+ + e \tag{25}$$

Also N_2^+ is formed, but reacts rapidly:

$$N_2^+ + O_2 = N_2 + O_2^+$$

$$N_2^+ + O = NO^+ + N \tag{26}$$

The main resulting ions in this region are O_2^+ and NO^+ (cf. Table 1 and Figure 8).

The main loss processes are, as in the D region, the reactions of dissociative recombination.

The region is in photochemical equilibrium during daytime. Motions and diffusion are also important at night.

F Region. There is no chemical distinction between the F_1 and F_2 regions. As seen in Table 1, atomic ions predominate.

Radiation in the interval 10–80 nm is primarily responsible for the ionization of the dominant neutrals O and N_2. O^+, N_2^+, N^+ are produced and, by subsequent reactions, O_2^+ and NO^+. O^+ becomes the dominant ion, while N_2^+ virtually disappears by reaction (see Table 1 and Figure 8).

The principal loss processes are again those of dissociative recombination for molecular ions. For atomic ions, such as the dominant O^+, the charge must first be transferred to a diatomic molecule (first of (21) followed by the first of (23)).

Reaction rates (both of photochemical production of ions and of dissociative recombination) become so slow in the upper regions due to the low concentrations, that the mean life of ions is long enough to allow them to distribute by diffusion. Thus, above ~ 300 km, diffusion becomes the controlling process for vertical distribution of ion densities. Electrons are also exponentially distributed above the maximum concentration in the F_2 region, again at about 300 km. At lower levels, the production of ions becomes more important. The formation of the electron density maximum at the peak of the F_2 region is due to the competition between the production and the diffusion processes.

6. Ozone in the Stratosphere (Ozonosphere)

Most of the UV radiation from the Sun lies in the region close to the visible, say between 170 and 370 nm (cf. Figure 3). This is the region absorbed by O_3. Should the O_3 layer be destroyed, this radiation would reach the ground, and possibly make life impossible. This is the reason for the present concern of atmospheric physicists that N oxides from combustion in supersonic jets flying in the stratosphere, or fluorinated compounds ('freons') arriving to the stratosphere due to their chemical stability, could damage the O_3 layer.

The distribution of O_3 can be measured from the ground or through 'ozonosondes'. The methods are spectroscopic, using its absorption properties in the UV or in the infrared, and chemical. Figure 9 illustrates the type of vertical distribution obtained for high and low latitudes. The units of concentration used here in abscissae are cm at STP per km, i.e. the thickness of ozone O_3, in cm, that would be obtained if all the O_3 contained in one km were concentrated pure and measured at standard temperature and pressure (273 K

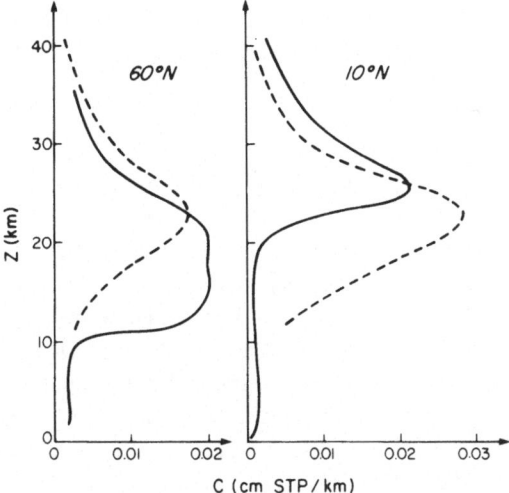

Fig. III-9. *Vertical distribution of ozone.* The full curves show typical distributions, as determined experimentally, for high (60°N) and low (10°N) latitudes. The dashed curves are typical of the expected concentrations, from the theory of photochemical equilibrium. z = altitude, c = ozone concentration expressed in thickness (cm) of pure ozone measured at standard temperature and pressure (STP = 273 K and 1 atm) per km of height.

and 1 atm). It may be remarked that the total amount of O_3 in the stratosphere is only in the range of 0.15 to 0.4 cm at STP.

As O_3 absorbs very strongly the UV, the upper layers of the ozonosphere already stop most of it; the absorption produces the increase in temperature in the stratosphere and the maximum of $\sim 0°C$ at the stratopause.

The formation of the O_3 occurs photochemically, and can be studied through the different reactions intervening and the knowledge of their specific rates. The dominant reactions are

coefficients

$$\begin{array}{lll}
\text{Primary} & \left\{ \begin{array}{ll} O_2 + h\nu = 2O & (\lambda < 242 \, nm) \\ O_3 + h\nu = O_2 + O & (\lambda < 1100 \, nm) \end{array} \right. & \begin{array}{l} f_1 \\ f_2 \end{array} \\
\\
\text{Secondary} & \left\{ \begin{array}{ll} O + O_2 + M = O_3 + M & \\ O + O_3 = 2O_2 & \end{array} \right. & \begin{array}{l} k_3 \\ k_4 \end{array}
\end{array} \qquad (27)$$

where M is a non-reacting molecule which acts as a third body carrying away the excess energy. Other reactions occur as well, but can be neglected in first approximation[*]. The velocities with which these reactions occur are proportional to the concentrations of the reactants. The proportionality coefficients (specific reaction rates) are known for the last two (k_3, k_4) and can be estimated for the first two for which these coefficients f_1, f_2 depend on the intensity of incident radiation and its distribution over λ. By assuming a

---

[*] E.g., $O + O + M = O_2 + M$, which occurs at low rate because of the scarcity of atomic O at these heights, $2O_3 = 3O_2$, etc.

steady state ('photochemical equilibrium'), it is possible to derive an expression for the concentration of O_3 as a function of height; the theory gives

$$[O_3] = \left[\frac{f_1 k_3}{f_2 k_4} r\right]^{1/2} \cdot [O_2]^{3/2} \tag{28}$$

Here f_1, f_2 are functions of the height, and so is $[O_2]$ (the bracket indicating concentration). $r = [M]/[O_2]$.

This profile of concentration can be compared with the measurements. Typical curves are shown in Figure 9 (dashed lines). It may be seen that less O_3 than theoretically predicted is found over the tropics (with higher maximum), while more (and with lower maximum) is found in high latitudes. These differences suggest that O_3 is formed more actively in low latitudes (stronger radiation) and transported meridionally towards high latitudes, where it penetrates to the troposphere through the jet discontinuities (cf. Ch.VII, §10) and eventually reaches the ground where it is destroyed (see sketch in Figure 10). This is consistent with the conclusions derived from the low concentrations of water vapour in the stratosphere (cf. Ch.II, §4).

In fact, the assumption of photochemical equilibrium is only valid above 30 km, where the two primary photochemical reactions occur rapidly. At lower altitudes, the active radiation disappears rapidly, with the result that O_3 becomes stable and can only be consumed by chemical reactions at the ground, where, being a strong oxidant indeed, it will quickly react with oxidizable materials.

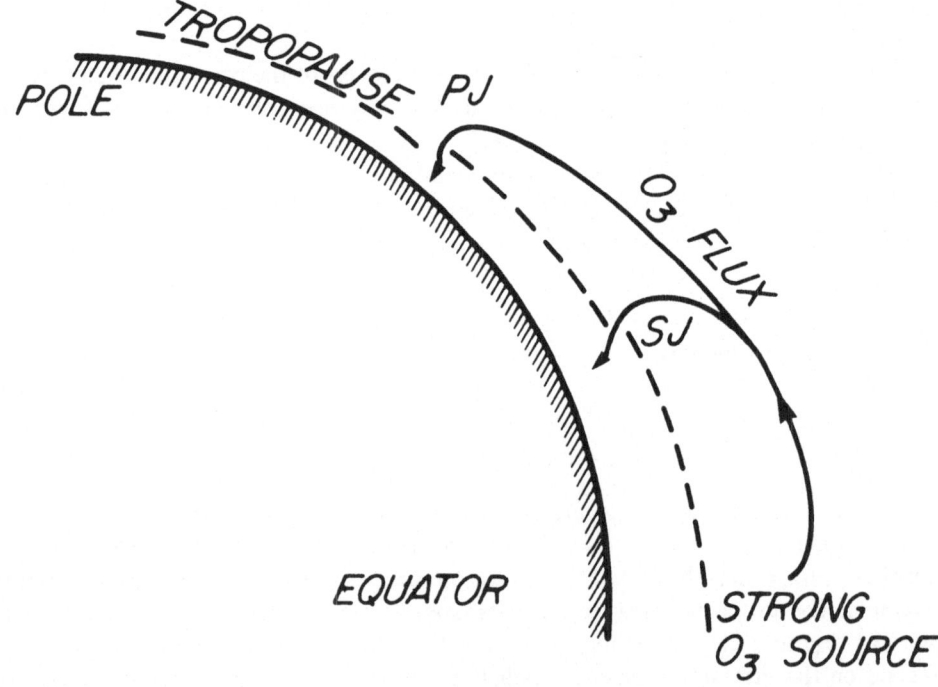

Fig. III-10. *Meridional transport of ozone.* The strongest source region is above the tropics. O_3 produced there is transported meridionally and penetrates into the troposphere through turbulent mixing at the subtropical and polar jets (SJ and PJ) (see Ch. VII, §10). Cf. Figure II, 4.

(c) Terrestrial Radiation

7. The Greenhouse Effect

As we have seen in §2 and Figure 3, the long wave radiation corresponding to the terrestrial temperatures covers the range of 4 micrometers to several tens of micrometers, and can be considered separately from the short wave or solar radiation. The first concern, in considering the atmosphere, should be to study its absorption properties in this range, in order to derive later the consequences of these properties. However, it will be easier to treat first a simpler case and then to extend our analysis to the atmosphere.

Let us consider a simplified example of the conditions in a greenhouse. We shall make the three following simplifying assumptions:
1) There is no heat loss from the ground to the air (we shall see that this is an oversimplification);
2) The rate of heat loss from the surface layers of the ground to deeper layers can be neglected;
3) The influx of long wave radiation from the atmosphere down to the greenhouse is negligible when compared with the influx of solar radiation, and can therefore be neglected.

We start by considering a portion of the ground, receiving solar radiation, of which a large fraction is absorbed; let us call E this fraction of the energy flux referred to the unit area of ground (another fraction will be reflected, according to the particular properties of the ground, but we will not be concerned with this part). For the moment we are not assuming that there is any roof above the ground. The surface layers will become hotter due to the absorbed energy, and its radiant emittance will increase accordingly. As the ground, as an emitter, can be assumed to act as a black body, its integral emittance will be σT_0^4, T_0 being its temperature. This temperature will increase until the emitted energy balances the incoming solar energy; a radiative equilibrium is then reached, expressed by:

$$E = \sigma T_0^4 \tag{29}$$

The situation is depicted in Figure 11a.

Now let us assume that at a few meters above the ground a glass roof is installed, as shown in Figure 11b. The radiant equilibrium will change drastically. Now the glass lets the solar radiation through (also the reflected fraction will go back unimpeded), but stops altogether (as a black body) the long wave radiation emitted by the ground and emits energy itself from its two surfaces, in the amount of σT^4, upwards to the atmosphere and also downwards towards the ground; T is the glass temperature, which we consider uniform throughout its thickness. The ground is then receiving more energy than before, and its temperature will therefore increase until an equilibrium is reestablished, in which both the ground and the glass emit as much as they absorb. Let T_0' be the final temperature of the ground.

In the new steady state, the upward emission of the glass, which is the only energy going back to the atmosphere, must again balance the incoming energy E. Therefore

$$E = \sigma T^4$$

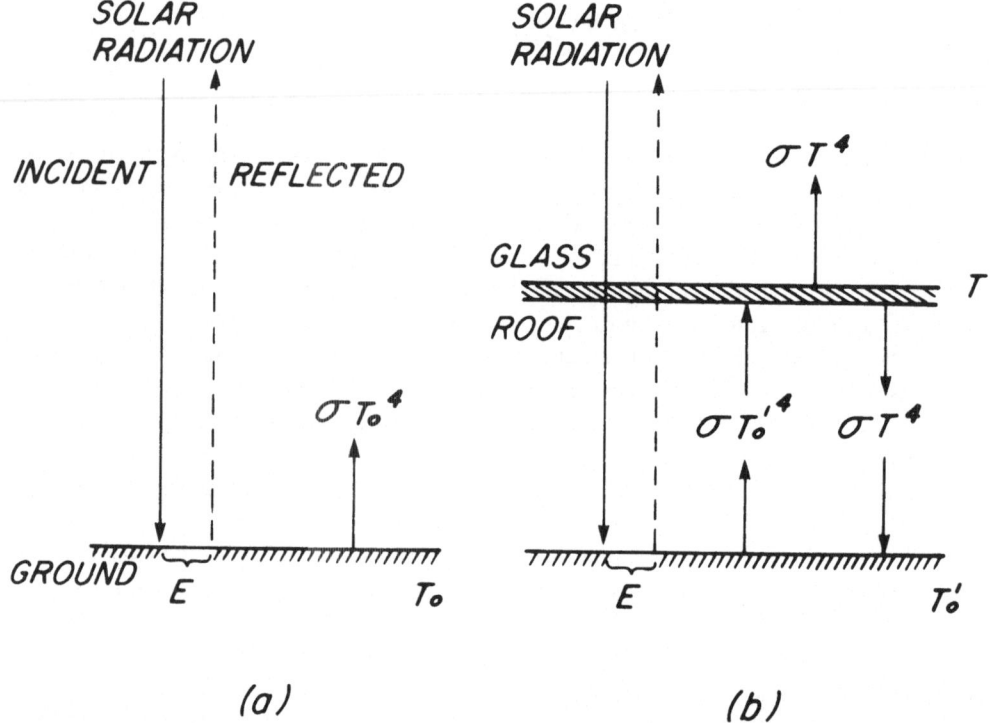

Fig. III-11. *Greenhouse effect.* (*a*) Balance of radiation fluxes without glass roof; (*b*) same, with glass roof. E = solar energy absorbed by the ground, per unit area and unit time. T_0, T_0' = ground temperature, without and with glass roof, respectively. T = glass roof temperature.

and comparing with (29),

$$T = T_0$$

The glass has now the temperature that the ground had before introducing the roof. The balance of energy per unit area of glass roof will be

$$2\sigma T_0^4 = \sigma T_0'^4$$

and therefore

$$T_0' = 2^{1/4} T_0 = 1.19 T_0$$

For instance, if $T_0 = 293\,\mathrm{K} = 20°\mathrm{C}$, $T_0' = 348\,\mathrm{K} = 75°\mathrm{C}$. This result should not be taken literally, in view of the simplifying assumptions. Nevertheless we see that the glass roof is capable of producing a considerable increase in the temperature of the ground and, consequently, in the air within the greenhouse, and that this effect is due to the trapping of radiant energy, once it has been transformed from short wave to long wave.

In fact, the effect of the glass cover in greenhouses depends as much, or more, on its preventing heat losses through air convection as from the radiation effect just described. It is the latter, however, that is referred to when the expression 'greenhouse effect' is used, and its importance transcends the simple application to greenhouses in that it also applies to the general effect of the presence of the atmosphere on the temperature of the

Earth. The problem becomes more complicated because the atmosphere is not a black body in long wave, as is the glass roof. We shall consider it now.

8. Emission and Absorption of 'Long wave' (Terrestrial) Radiation

If we go back to the schematic drawing of Figure 1 and take into account the wavelength range covered by terrestrial radiation (Figure 3), we see that emission and absorption of this radiation must be associated with transitions of rotational and vibrational states only.

Now we consider that in this range:

1) The ground surface behaves approximately as a black body. Its emissivity may actually be in the range of 0.90–0.95, but for the sake of simplicity in our arguments, we will take this as approximately unity;

2) Liquid water has an absorption spectrum similar to water vapour (see below), but with a much higher coefficient of absorption. Thus, any cloud layer of thickness > 100 m, with 200 droplets/cm^3 of 10 μm radius, behaves as a black body, i.e. practically any cloud is a black body;

3) Air is only semi-transparent.

In order to understand what happens to the radiation emitted by the ground, we must therefore know the absorption properties of the atmospheric air. Let us consider now this problem.

The major constituents, N_2 and O_2, do not absorb in this region. The absorption properties are due to the presence of water vapour, carbon dioxide and ozone. This is schematically indicated in Figure 3. The main facts are:

$Water\ Vapour$ — Absorbs in $\left\{ \begin{array}{l} - \text{some bands at} < 4\,\mu\text{m} \\ - \text{intense band at } 6.3\,\mu\text{m} \\ - \text{strong broad band starting} \\ \ \ \ (\text{weakly}) \text{ at } 9\,\mu\text{m and} \\ \ \ \ \text{increasing for longer } \lambda. \end{array} \right.$

CO_2 — Absorbs in band at 13–17 μm.
O_3 — Absorbs in intense narrow band at 9.7 μm.

If we disregard the ozone, which has interest only with respect to the ozone layer in the stratosphere and only absorbs in a narrow interval, the result is that, due to H_2O and CO_2, air absorbs in most of the terrestrial spectrum, but a region remains around 9–11 μm, for which it is practically transparent: the *atmospheric window*.

Figure 12 summarizes the situation, giving the emission spectrum of a thin layer of atmospheric air at 300 K. By a 'thin layer' we understand here a layer containing 0.3 mm of precipitable water in the form of water vapour*; such a layer could have a thickness of the order of 200 m close to the ground, considerably more in higher layers, which contain much less humidity. The radiant emittance E_λ of this layer is plotted in the vertical axis in relative units. The emission spectrum of a black body at the same temperature is plotted for comparison. It can be seen that the thickness of a 'thin layer' has been chosen

---

* I.e. an amount of water vapour such that, if condensed into liquid water, would have a thickness of 0.3 mm.

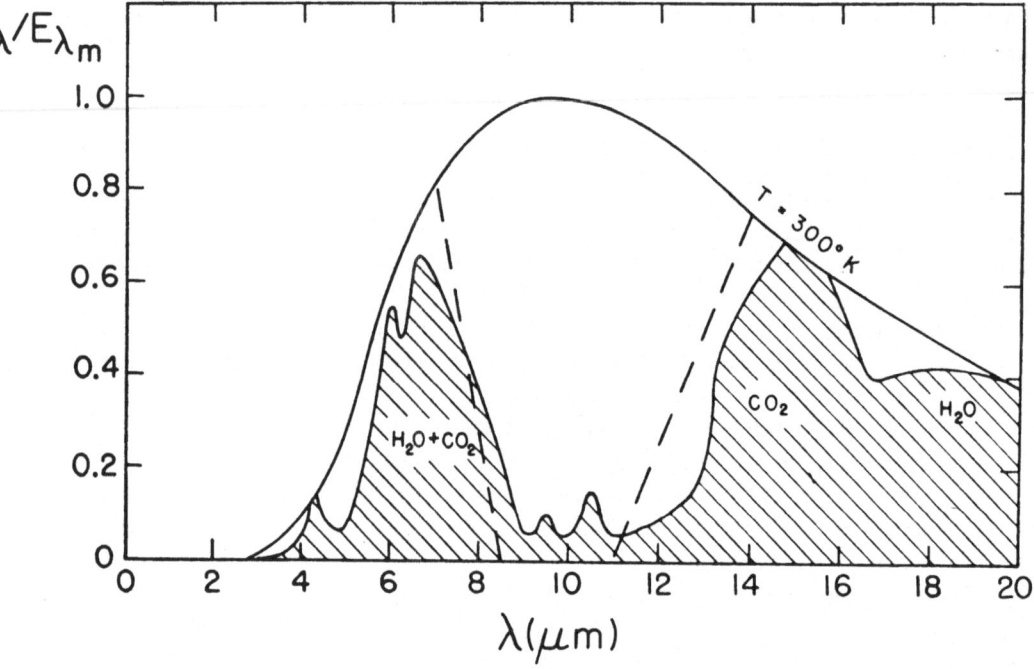

Fig. III-12. *The emission spectrum of the atmospheric gases for a thin layer of atmosphere at sea level.* E_λ = radiant emittance; E_{λ_m} = maximum radiant emittance at 300 K; λ = wavelength. The shaded area is the emission spectrum, whereas the envelope marked $T = 300$ K is that of a black-body curve at that temperature. Since the area enclosed by the gases is less than that of a black body at the same temperature, it follows that the gases absorb and emit less energy than a black body of the same temperature. The dashed lines are the limits of a simplified emission spectrum for water vapour and carbon dioxide. (Copyright © 1954 by The Massachusetts Institute of Technology.)

so that in the two main shaded regions the layer acts (in both emission and absorption, according to Kirchhoff's law) approximately as a black body; a thinner layer would let some of this radiation go through, and a thicker one would have emitted, absorbed and re-emitted within the same layer.

We shall now refer briefly to an approximate treatment of the long wave radiation transfer through the atmosphere due to Simpson. Let us imagine the atmosphere divided in a number of 'thin layers', and let us find out how much energy the Earth will be losing to space through its air envelope. This will be illustrated by Figure 13. We recall again that, according to Kirchhoff's law, the emissivity of any object is equal to its absorptivity; therefore the atmosphere will be opaque (it will absorb all radiation) in the intervals of λ in which it emits as a black body, and will be transparent where it does not emit anything. Where it emits less than a black body, i.e. where its absorptivity has some intermediate value between 0 and unity, it will also absorb partially.

Assume now that the ground is at 14°C. Its black body emission will go undisturbed through the whole atmosphere only in the window region, say between 8.5 and 11 μm (transparent region). This loss is indicated by the corresponding column in the figure. In the regions where 'thin layers' (as defined before) absorb like black bodies, i.e. for $\lambda < 7 \mu$m and $\lambda > 14 \mu$m (opaque regions), the energy irradiated by the ground will be

68

absorbed by the layer just above it; this layer is also emitting from its top, but this energy is absorbed by the next layer; and so on, until we arrive to the last layer containing sufficient water vapour and CO_2. This last layer will have its top around the tropopause, with a temperature which is here assumed to be $-60°C$. In these wavelength regions, therefore, the atmosphere is losing energy like a black body at $-60°C$, as indicated in the figure. In the intermediate regions, where the 'thin layers' are semi-transparent (rather than transparent or opaque), the energy lost will be somewhere between the emission of a black body at 14°C and a black body at $-60°C$. This is indicated in the figure by the curves for $7\,\mu m < \lambda < 8.5\,\mu m$ and $11\,\mu m < \lambda < 14\,\mu m$. The total energy lost through the atmosphere is then given by the shaded area in the figure.

When seen from space, the Earth appears therefore (in long wave) quite different from a black body. Its emission curve looks like the envelope of the shaded area in Figure 13 rather than like a Planck curve. As for the greenhouse effect, it is obvious that it applies to the atmosphere in spite of the fact that it is only partially opaque to long-wave radiation. If there were no atmosphere over the Earth, the emission would be of the black body type, and a much lower ground temperature would suffice to balance the incoming solar radiation. The fact that there is an atmosphere, transparent to short-wave but partly trapping the outgoing long-wave radiation, makes necessary higher ground temperatures in order to keep a radiative equilibrium.

The previous treatment is only approximate, but it does pose the basic ideas involved in the transfer of radiant energy through the atmosphere. A rigorous treatment is possible, starting from differential equations for the absorption and emission from infinitesimal

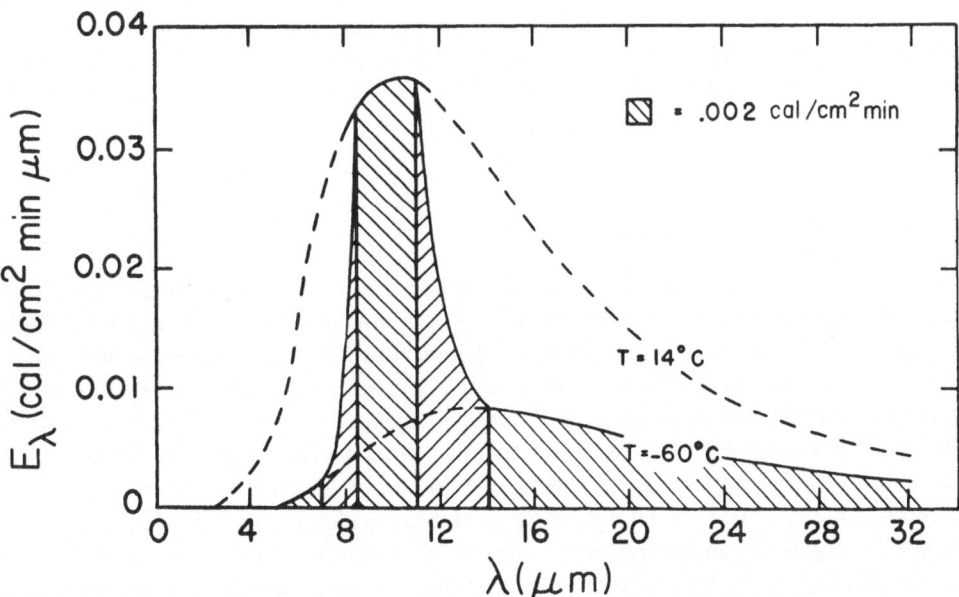

Fig. III-13. *Long wave energy lost by the Earth to space.* E_λ = radiant emittance; λ = wavelength. The shaded area gives the approximate estimate, assuming that 14°C is the ground temperature and $-60°C$ the temperature of the last 'thin layer'. The small square area gives the energy equivalence. (Copyright © 1954 by The Massachusetts Institute of Technology.)

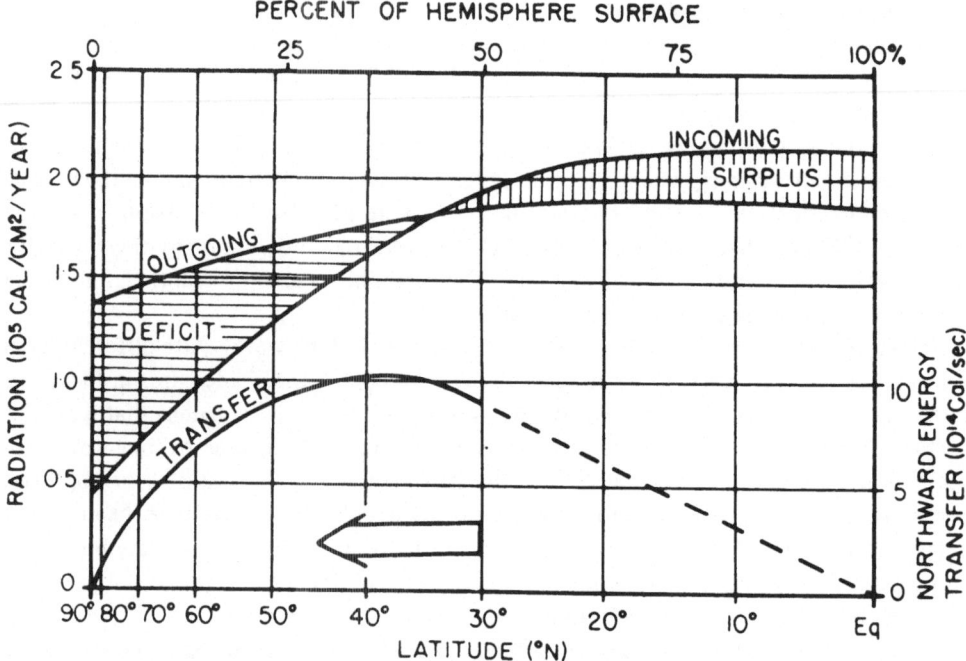

Fig. III-14. *Meridional transport of energy*. The imbalance between radiant energy absorbed and emitted – surplus in low latitudes and deficit in high latitudes – must be compensated by northward transfer in the atmosphere. (Copyright © 1964 by Scientific American, Inc.)

layers rather than the 'thin layers' defined above (Schwarzschild equations). Their integration is made difficult by the rapid variability of absorption as a function of wavelength; the problem can be solved, however, with adequate simplifications, and the results can be applied by using 'radiation charts'.

9. Other Consequences of the Greenhouse Effect

The greenhouse effect must also be taken into account in certain situations of great practical importance. During the night, when there is no incoming solar radiation but the emission in long wave continues, the ground cools down by many degrees of temperature and, by conduction, so do the adjacent air layers. In fact, this usually produces a temperature inversion just above the ground. At certain times of the year, the cooling results in the production of *frost*, with its consequent damage to certain crops. The danger of frost decreases with increasing air humidity, due to the increased absorption of the lower layers. A much better protection is the presence of clouds, because clouds are black bodies in long wave, and therefore act like the glass roof of a greenhouse. The lower the cloud, the higher the temperature of its base and the better the protection. Thus frosts will develop most likely during the night (in otherwise similar starting conditions) when the sky is clear and the air dry; least likely for low cloud cover and high humidity. Again, clouds and (less efficiently) water vapour act as a blanket in preventing high radiation losses.

70

According to these ideas, some of the methods for frost prevention are based on obtaining the same effect of radiation protection by artificial means: a fog or a smoke blanket is produced, covering the ground. Artificial ventilation is also used as a means to increase heat exchange with the air and prevent the temperature inversion.

Still another example of the greenhouse effect is given by the possible consequences of a substantial increase in the concentration of one of the absorbants in the atmosphere: carbon dioxide. The huge development of industries has resulted in the introduction of this gas in such large amounts that its concentration is rising at the approximate rate of 0.8 ppm per year (the total concentration being 330 ppm). If this trend continues or is accentuated, the consequence may be an increase in the average temperature of the Earth, leading to enhanced biological activity, increased melting of ice caps, etc.

10. Energy Budget

We have mentioned before that the Earth, being on average in a steady state with a constant average temperature, must lose to space as much energy in long wave radiation as it receives from the Sun in short wave radiation. The energy received from the Sun is characterized by the energy flux of solar radiation incident over a perpendicular unit area outside the atmosphere; this is called the *solar constant S*, and is equal to (1.94 cal/cm^2.min = 1353 W/m^2) $\pm$ 1.6%. Therefore the total power being received by the Earth from the Sun is equal to this value multiplied by the cross section of the Earth $\pi R^2 = 1.27 \times 10^{14}$ m^2. If we average this power over the whole surface of the Earth, we see that the power received per unit surface area is

$$S \times \frac{\pi R^2}{4\pi R^2} = 0.49 \text{ cal/cm}^2.\text{min} = 338 \text{ W/m}^2 \tag{30}$$

This same amount must be lost to space in long wave radiation.

Obviously, the solar radiation is not received uniformly over the Earth, neither is the terrestrial loss uniformly distributed. For instance, the equator receives on the average several times more annual solar energy than the poles; the long wave loss is also larger, but less markedly. It follows that there is a surplus energy for the lower latitudes (< 30–$35°$) and a deficit for the higher latitudes (> 30–$35°$). This is illustrated in Figure 14. As the tropics do not get progressively hotter or the higher latitudes colder, a meridional transport of energy must continually occur, from lower to higher latitudes, in order to compensate the imbalance. This is important in connection with atmospheric motions as will be seen in Ch.VII.

Here we shall only consider the heat budget in the average for the whole planet during the year.

Not all the solar radiation incident over the Earth is absorbed by the atmosphere and the ground. A large fraction of it is reflected from the tops of the clouds or from the ground. Given a surface, the fraction of solar radiation reflected (in the given conditions) is called the *albedo* of that surface. An average planetary albedo can be estimated for the whole Earth: this is 0.31 = 31%. Let us consider the schematic Figure 15. Here 100 arbitrary units are assigned to the power received per unit area as solar radiation (0.5 cal/cm^2.min), and all the figures are referred to these units. We should remark that the planet-

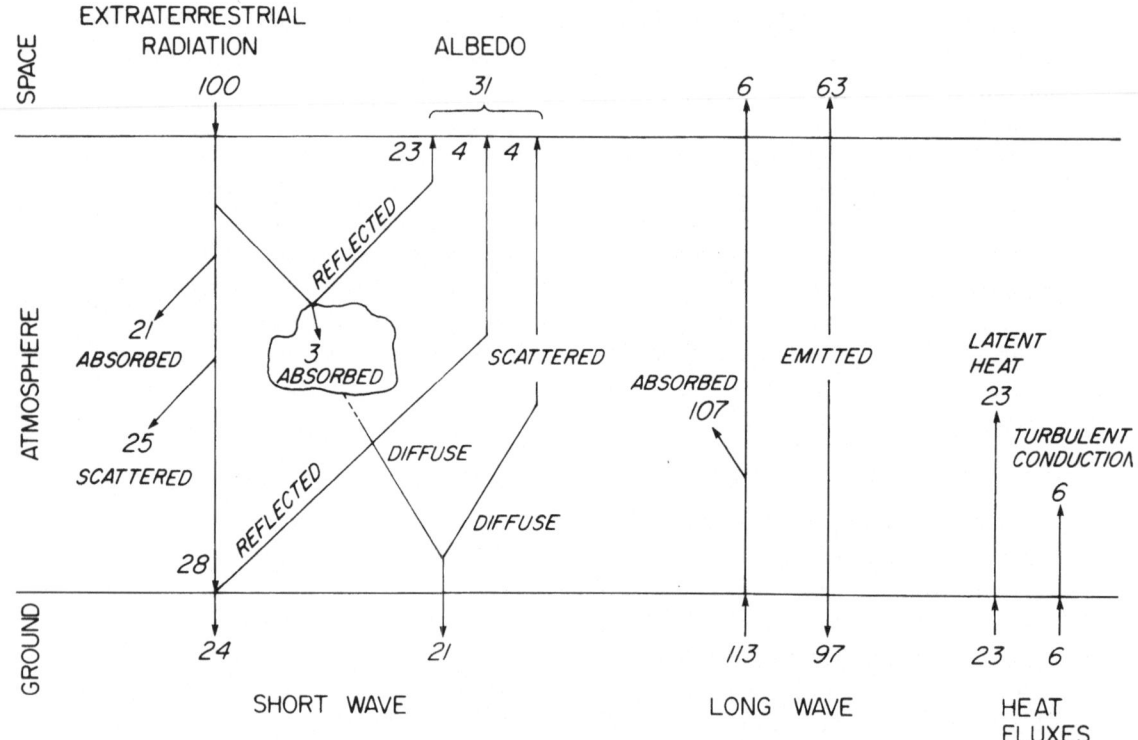

Fig. III-15. *Energy budget.* Figures are in arbitrary units referred to 100 for the oncoming solar radi-ation. This estimated budget is an average for the whole Earth during the year. (Figures based on Barry and Chorley, see Bibliography.)

ary albedo includes not only direct reflection but also the fraction of radiation which has been scattered by the air molecules and particles and has found its way back to space.

The 100 units of incoming solar radiation become distributed as follows. 21 units are absorbed (17 in the troposphere and 4 in the stratosphere); 26 fall on clouds, which absorb 3 and reflect 23 toward space. 25 units are scattered mainly by the air molecules (Rayleigh scattering) and also by clouds. 28 directly reach the ground, which absorbs 24 and reflects 4 towards space. The scattered light constitutes the diffuse luminosity of the sky during the day; this luminosity is predominantly a blue colour because the shorter wavelengths are scattered more than the longer ones. Together with some diffuse radiation from clouds (included in the 25 units computed as scattered), the amount reaching the ground and being absorbed by it is about 21 units, while 4 units are scattered back to space.

The 45 units of solar radiation energy absorbed by the ground are compensated by losses from the ground as long wave radiation and heat fluxes. An estimate of 113 units (of the same value as before) are emitted by the ground in infrared radiation, according to the black body emission at 288 K. Of this, 6 units escape directly to space through the atmospheric window, and the rest (107) is absorbed in the atmosphere. The atmosphere itself radiates in long wave, 63 units of which are lost to space, while 97 fall on and are absorbed by the ground.

72

The additional heat fluxes that transport energy from the ground to the atmosphere are the latent heat of water evaporation (23 units) and conduction by turbulent flow (6 units).

These estimated gains and losses must compensate in the average steady state. Thus for the atmosphere we have:

	Gains		Losses
Short wave absorption		Short wave scattered	
(by air and clouds)	24	to space	4
Scattering	25	Short wave scattered	
Long wave absorption	107	to ground	21
Latent heat	23	Long wave emitted to	
Turbulent conduction		space	63
from the ground	6	Long wave emitted to	
		ground	97
Total gain	185	Total loss	185

Similarly, the 100 units of solar radiation which constitute the total input of energy coming from space are compensated by a global reflection of 31 (albedo), made up of reflection from clouds and ground and back scattering, and 69 units lost to space as infra-red radiation. It may be easily checked that the gains and losses also compensate when computed for the ground.

The figures mentioned should be taken only as orientation estimates, since many of them can obviously only be obtained by rather crude computations. However, the general picture is undoubtedly correct.

11. Effective Temperature of the Earth and Other Planets

The *effective temperature* of a planet (or of the Sun) is defined as the temperature that it should have if, behaving as a black body, it would irradiate the same amount of energy per unit time. In that case, the total power lost by radiation should be equal to the surface area A of the planet times σT_e^4, where T_e = effective temperature, according to (5). We have seen in the previous section that, taken as a whole, as seen from space, the Earth irradiates in quite a different way (cf. Figure 13). However, the effective temperature gives an indication of the range of the planetary temperatures.

The effective temperature of the Earth can be derived easily by considering that the solar energy received must be equal to the energy lost to space. Thus we may write that

$$S(1 - \alpha)\pi R^2 \ = \ \sigma T_e^4 4\pi R^2$$

where S = solar constant, α = planetary albedo and R = Earth's radius. From this equation, T_e can be readily derived as:

$$T_e \ = \ \left[\frac{S(1 - \alpha)}{4\sigma} \right]^{1/4} \tag{31}$$

The effective temperatures of the different planets are given in Table 2:[*]

[*] After Goody and Walker. See Bibliography.

TABLE 2

Effective temperatures

Planet	Distance from Sun (10^6 km)	Flux of solar radiation (J cm^{-2} sec^{-1})	Albedo	T_e(K)
Mercury	58	0.92	0.058	442
Venus	108	0.26	0.71	244
Earth	150	0.14	0.33	253
Mars	228	0.060	0.17	216
Jupiter	778	0.0049	0.73	87
Saturn	1430	0.0015	0.76	63
Uranus	2870	0.000 37	0.93	33
Neptune	4500	0.000 15	0.84	32
Pluto	5900	0.000 089	0.14	43

Chapter III: Questions

Q1. Of two platinum strips, one is polished and the other is covered with platinum black (finely divided platinum), which increases its absorptivity in the visible. Both strips are heated to 1000°C; which will appear brighter?

Q2. Is 'long wave' radiation, emitted by the ground, capable of producing chemical reactions in the atmosphere? Is it capable of increasing the rotation energy of some molecules?

Q3. Why is there a maximum of temperature at the stratopause?

Q4. Why are oxygen atoms not produced in the troposphere and why is there a maximum concentration [O] at around 100 km of altitude?

Q5. Why is the ozone layer important for the preservation of life at the ground?

Q6. Above about 1500 km, H and He are the predominant gases in the exosphere. Explain why.

Q7. Why are there no negative ions in the upper ionosphere?

Q8. Why do attachment reactions (such as recombination of a positive ion with an electron) become less important with increasing height?

Q9. The colours of stars are related to their temperature, whereas the colours of the planets are not. Explain.

Q10. Assume that the 'thin layers' defined in §8 started to behave like black bodies in long wave radiation. What would happen with the temperature of the tropopause (in average), and how would it compare with the effective temperature of the Earth?

Q11. Low clouds emit more infrared radiation than high clouds of comparable thickness. Explain.

Q12. Temperature inversions tend to form at night immediately above the tops of cloud layers. Explain.

Q13. The effective temperature of the Earth is lower than the mean temperature of the Earth's surface by about 30 K. Explain.

Q14. The effective temperature of Venus is lower than that of the Earth, even though Venus is nearer to the Sun. Explain.

Q15. Production of smoke is frequently used in agriculture to prevent frost danger. Explain.

Chapter III: Problems

(Any necessary constants not given in the statement of a problem will be found in the Table of Constants on pages x–xi)

P1. Consider radiation of wavelength $\lambda = 230$ nm. Calculate:
(a) its frequency ν;
(b) the energy of one photon;
(c) the energy per mole.
(d) Is this radiation capable of dissociating the O_2 molecule? The corresponding dissociation energy is 494 kJ/mol.

P2. Calculate the power irradiated by 10 cm^2 of surface of a black body at 500°C.

P3. Consider Figure 2. (a) How much radiant power will a black body at 6000 K emit in the interval of λ between 500 and 550 nm? The interval of λ can be approximately considered as a differential. Check whether the result agrees with the figure. (b) What would be the answer for a body with absorptivity $A_\lambda = 0.3$ in that interval?

P4. Derive Wien's displacement law $\lambda_m \cdot T = \text{const}$ from Planck's law.

P5. If the maximum of $E_\lambda = f(\lambda)$ occurs at $\lambda = 500$ nm for the solar radiation, and the temperature of the Sun is ~ 6000 K, where will be the maximum of E_λ for terrestrial radiation from ground at 300 K?

P6. Knowing that the solar constant S (solar radiation received outside the atmosphere per unit time and unit normal area) is 2.0 cal/cm^2.min, and that the solar radius is 6.9 × 10^8 m, derive the effective temperature of the Sun.

P7. Solar radiation is reaching the ground at a rate of 0.9 cal/cm^2.min. The ground considered has an integral absorptivity A of 0.2 in short wave (solar radiation) and 1 in long wave (terrestrial radiation). (a) What will be the radiative equilibrium temperature of the ground surface? (b) What will be the steady state temperature if half of the absorbed energy is lost by conduction to the air and to deeper layers of the ground?

P8. Assume that n glass roofs are placed above the ground and one above the other, and that radiative equilibrium is reached after a time. Assume that there is no heat conduction through the air, or downwards from the surface layers of the ground to deeper layers. Solar radiation is received by the top glass plate, with a flux density E. Radiation from the atmosphere above the top plate is negligible. The ground has 0 albedo and no reflection occurs on the glass plates.

(a) Calculate the temperature of each plate and of the ground, in terms of the temperature of the top plate. Number the plates starting from the top (with temperature T_1). Use T_0 for the temperature of the ground.

76

(*b*) What is the value of T_0, if $n = 5$ and $E = 0.5\,\text{cal/min}.\text{cm}^2$?

P9. At a certain time (at noon, during an equinox) and longitude, the Sun is at the zenith over the equator.
   (*a*) Compare (give the ratio) of the incident radiation at the top of the atmosphere, per unit area, for the equator and for $60°$ of latitude North over the same meridian.
   (*b*) If the ground temperatures are $310\,\text{K}$ at the equator and $270\,\text{K}$ at $60°$ North, give the ratio of the ground emissions.

P10. The distance between Sun and Earth varies during the year. It is minimum in January and about 3.3% larger for the maximum in July. What will be the corresponding seasonal change in the effective temperature T_e?

IV. Atmospheric Thermodynamics and Vertical Stability

1. Atmospheric Systems

From the thermodynamical point of view, a *system* is any body of a given mass and composition under study. All the rest of the bodies with which the system may eventually interact are called its *surroundings*.

In the atmosphere we deal basically with two types of systems: those composed of air, with a humidity that can go from 0 to saturation, and clouds, composed of saturated air and either water droplets or ice crystals. (There are also mixed clouds, with both drops and crystals, but we shall not consider them.) We shall be dealing in this chapter exclusively with the troposphere.

Unsaturated air can be considered as a mixture of:

a) Dry air: a mixture of gases in constant composition, as seen in Ch.II, with mean molecular weight 28.964.

b) Water vapour: its content can be expressed, for instance, by the vapour pressure or, as we shall do here, by its molar fraction N_v, defined as

$$N_v = \frac{\text{number of moles of water vapour}}{\text{total number of moles}} \simeq \frac{\text{number of moles of water vapour}}{\text{number of moles of dry air}} \tag{1}$$

where the last approximate equality results from the fact that the number of moles of water vapour is always much smaller than that of dry air.

Both dry air and water vapour in atmospheric conditions, and also their mixture as moist air, can be considered with good approximation as ideal gases. The presence of water vapour alters slightly the average molecular weight of the mixture, but with enough approximation for our purposes, we can assume that the average molecular weight of the mixture is always $M = 28.9^*$ and that the equation of state is the corresponding ideal gas law:

$$pV = \frac{mRT}{M} = nRT \tag{2}$$

where p = pressure, V = volume, m = mass, $R = 8.3$ J/mol.K = universal gas constant, $M = 28.9$ = average molecular weight and n = total number of moles $\simeq$ number of moles of dry air.

The presence of water vapour will also change slightly other parameters, such as the heat capacity of the air, but we shall ignore these corrections.

When we deal with clouds, the presence of water droplets or ice crystals also alters some parameters. However, the content of condensed water is only of the order of one or

---

* For air saturated with water vapour at 30°C, a rather extreme condition, the average molecular weight turns out to be 28.5.

a few grams per kilogram of air, and therefore its influence can be neglected, except for one important fact: latent heat is released or absorbed when water changes its physical state, i.e. during evaporation, condensation, freezing or melting. This important effect must be taken into account in the formulas.

2. First Principle of Thermodynamics, as Applied to Air and Clouds

The general formulation of the First Principle of Thermodynamics is:

$$dU = \delta Q + \delta A \tag{3}$$

Here U = internal energy of the system, Q = heat absorbed by the system, A = work performed upon the system by external forces. U is a state function, i.e. its value is determined (except for an arbitrary additive constant) by the state of the system; therefore the change dU, or ΔU for a finite process, depends only on the initial and final states, and not on the path followed by the process. A and Q are not state functions, so that δA and δQ, or A and Q for finite processes, depend on the path followed by the process.

In the atmosphere, the only work that can be received by the system is expansion work; therefore:

$$\delta A = -p\, dV \tag{4}$$

Introducing (4) into (3), the first principle becomes

$$dU = \delta Q - p\, dV \tag{5}$$

The first principle can also be expressed in terms of another state function, called the *enthalpy H* and defined by

$$H = U + pV \tag{6}$$

If we differentiate (6) and introduce (5), we obtain

$$dH = \delta Q + V dp \tag{7}$$

which is the alternative expression for the first principle.

We shall now apply equation (7) to the systems of our interest: first to clear air, and then to clouds.

Let us consider one mole of air. We define the *molar heat capacity of air at constant pressure*, C_p, as the heat necessary to increase the temperature of one mole of air by one degree at constant pressure. Now let us assume that we heat one mole of air at constant pressure ($dp = 0$), producing a temperature change dT. Then (7) gives

$$(dH)_{p=\text{const}} = (\delta Q)_{p=\text{const}} = C_p\, dT \tag{8}$$

But it can be shown in thermodynamics that the enthalpy (as also the internal energy) of ideal gases is a function of only the temperature. Then the value of dH does not depend on whether the pressure is kept constant or not. Therefore we can write, for *any* (infinitesimal) process:

$$dH = C_p\, dT \tag{9}$$

80

Introducing (9) into (7) we obtain

$$C_p \, dT = \delta Q + V dp \tag{10}$$

This is the expression that we need to study the atmospheric processes of air.

Now let us consider one mole of saturated air containing water droplets (a cloud of water droplets; the case of ice crystals would be similar), and let us repeat the same arguments as before. When we consider a process at constant pressure, the heat received by air includes now two terms: one due to the increase in temperature, as before, and another one due to the latent heat absorbed by water if any evaporation takes place. We define the *molar latent heat of vaporization*, L_v, as the heat necessary to evaporate one mole of water. Then the new term that we must consider will be of the form $L_v \, dN_v$ where dN_v is the increase in molar fraction of water vapour in the air, i.e. approximately the number of moles of water evaporated in each mole of air. Thus, instead of Equation (8), we have now:

$$(dH)_{p=\text{const}} = (\delta Q)_{p=\text{const}} = C_p \, dT + L_v \, dN_v \tag{11}$$

We repeat now the previous argument about the enthalpy, to show that we can write, for any process:

$$dH = C_p \, dT + L_v \, dN_v \tag{12}$$

And, introducing this expression into (7):

$$C_p \, dT + L_v \, dN_v = \delta Q + V dp \tag{13}$$

This is the formula that we need in order to study atmospheric processes in a water cloud (a similar expression would be valid for an ice cloud, with appropriate substitutions for C_p and L_v).

dN_v can be expressed in terms of T and p as independent variables. According to the definition of N_v and to the gas laws,

$$N_v = \frac{e}{p} \tag{14}$$

where e is the partial vapour pressure of water and p the total pressure. Differentiating:

$$dN_v = \frac{de}{p} - \frac{e}{p^2} \, dp \tag{15}$$

This differential will only appear for the cases when the air is saturated with water vapour, so that dN_v results from the evaporation or condensation of water. In that case, de is linked to the variation of temperatures by the Clausius–Clapeyron equation (cf. Ch.II, §4):

$$\frac{de}{dT} = \frac{L_v e}{RT^2} \tag{16}$$

where e is now the saturation vapour pressure (which in Ch.II, §4 was designated by p_s). Introducing this relation into (15), we obtain

$$dN_v = \frac{L_v e}{RT^2 p} \, dT - \frac{e}{p^2} \, dp \tag{17}$$

81

which can be substituted into (13) in order to have the formula expressed in terms of the variations of T and p.

3. Main Processes in the Atmosphere

There are three main types of processes in which we shall be interested:

1) cooling or warming at constant pressure
2) expansion or compression
3) mixing

Cooling at constant pressure will refer to a process at a constant level (e.g. at ground), where the pressure variations are small enough to be ignored in this context. Expansion and compression are linked with vertical motions in the atmosphere, as we know that pressure varies with height.

We shall consider them in turn.

4. Cooling

Dew and *frost* form as a result of condensation or sublimation of water vapour on solid surfaces on the ground, which cool during the night, by radiation, to temperatures below those corresponding to saturation. The process can be understood on the diagram of

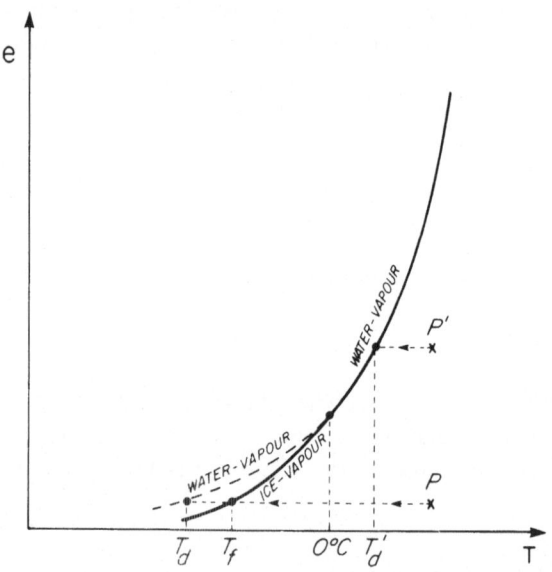

Fig. IV-1. *Dew point* (T_d) *and frost point* (T_f) *on a vapour pressure diagram.* (Schematic.) T = temperature; e = water vapour pressure. If air whose vapour pressure and temperature are represented by P is cooled isobarically, the temperature at which it reaches the ice-vapour equilibrium curve is the frost point, and the temperature at which it meets the water-vapour curve is the dew point. P' has only dew point.

82

saturation vapour pressures (cf. Ch.II, particularly Figure II,1) and is schematically presented in Figure 1. Cooling occurs at constant total pressure p (the pressure at ground). While the air is not saturated, the water vapour pressure, e, also remains constant, as $e = N_v \cdot p$ and N_v are constant. Thus if the air has temperature and water vapour corresponding to point P in the figure, cooling will proceed along a horizontal line towards the left, as indicated. When this line hits the ice-vapour equilibrium curve, the air becomes saturated with respect to ice and its temperature T_f is called the *frost point*. Actual freezing may not occur, as this process requires appropriate surfaces on which the vapour may sublimate. Then the cooling may proceed on to the water-vapour equilibrium curve, when air becomes saturated with respect to water; its temperature T_d is then called the *dew point*. Of course, air with humidities higher than that of the triple point, such as that represented by P' can only have the dew point T_d' (i.e. no frost can form by isobaric cooling).

Thus, if a mass of atmospheric air cools isobarically until its temperature falls below the dew point, condensation will occur as microscopic droplets formed on condensation nuclei (always present in the air; cf. Ch.V, §2); we call this a fog. It occurs due to the radiative cooling of the air itself or of the ground with which it is in contact (*radiation fogs*). As the droplets form, and because these droplets behave as black bodies in the wavelengths in which they irradiate, the radiation emitted by the layer increases, which favours further loss of heat. Condensation may also occur when an air mass moves horizontally over the ground toward colder regions, and becomes colder itself by turbulent heat conduction to the ground (*advection fogs*). In both cases, the cooling is practically isobaric, since pressure variations at the surface are usually very small.

Once condensation starts, the temperature drops much more slowly because the heat loss is partially compensated by the release of the latent heat of condensation. This sets

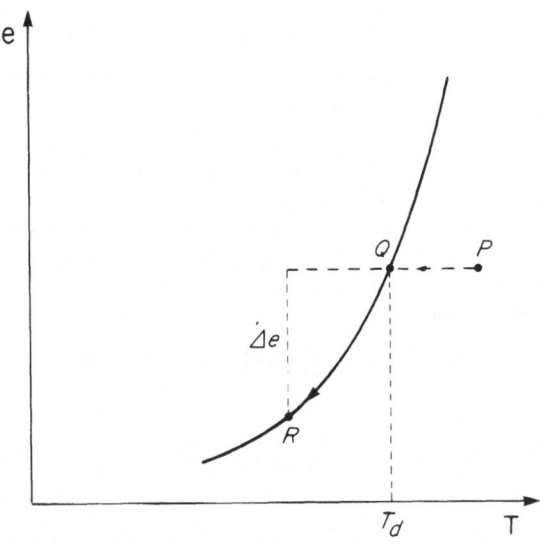

Fig. IV-2. *Isobaric cooling with condensation.* T = temperature; e = water vapour pressure; P = initial state of air; Q = saturation; R = final state of air; Δe = decrease of water vapour pressure due to condensation.

a virtual limit very close to the dew point, an important fact in forecasting minimum temperatures.

Let us consider again the process on a graph (Figure 2). From P to Q, and according to Equation (10), the simple equation

$$C_p \, dT = \delta Q \tag{18}$$

applies. From Q to R, water has been condensing, the water vapour pressure decreases by Δe, and we have a fog, i.e. a cloud at ground. Therefore Equation (13) applies, giving

$$C_p \, dT + L_v \, dN_v = \delta Q \tag{19}$$

In this equation, dN_v could be replaced by (17), giving the equation (notice that $dp = 0$):

$$\left(C_p + \frac{L_v^2 e}{RT^2 p} \right) dT = \delta Q \tag{20}$$

This means that the decrease in temperature could be related to the heat lost by the air δQ. δQ can be estimated for a given process (e.g., estimating the radiation loss during the night), and then the corresponding drop of temperature can be derived by applying (18) until the air saturates, and (20) after saturation.

5. Adiabatic Expansion Without Condensation. Potential Temperature

Vertical motions of air masses in the atmosphere are accompanied by changes in pressure. We have seen (Ch.I) that pressure decreases in an approximately exponential way with the altitude.

It is a good approximation to assume that these changes occur adiabatically, i.e. without exchange of heat with the surroundings: $\delta Q = 0$. The reason is that heat transfer processes from an air mass to its surroundings are slow in comparison with the air motion. These processes are energy transfer by radiation, and turbulent mixing at the borders. Mixing will not penetrate much in a short time, and what happens at the borders can be neglected if the air mass is large enough. Molecular heat conduction is entirely negligible.

Thus the process involved is the adiabatic expansion (or compression) of a gas. Application of (10) gives then:

$$\delta Q = C_p \, dT - V \, dp = 0 \tag{21}$$

By dividing this equation by T and applying the gas law ($V/T = R/p$), we obtain:

$$C_p \, d \ln T - R \, d \ln p = 0 \tag{22}$$

This can be readily integrated and manipulated to the form:

$$T p^{-R/C_p} = \text{const} \tag{23}$$

This is Poisson's equation for the adiabatic expansion of an ideal gas.

As air ascends in the atmosphere, therefore, T varies with p according to (23). If the air is taken adiabatically to the pressure $p_0 = 1000$ mb (used as a reference level), it will acquire, by definition, the *potential temperature* θ:

84

$$\theta = T \left(\frac{p_0}{p}\right)^{R/C_p} \tag{24}$$

θ is an invariant along the whole expansion and is used as a characteristic parameter of the air.

6. Adiabatic Expansion With Condensation

As the air rises, its vapour pressure e decreases with the total pressure, p, as

$$e = N_v p \tag{25}$$

where the molar fraction of vapour, N_v, is a constant. Simultaneously, T decreases according to (23). Since the saturation vapour pressure e_s is a rapidly-varying function of T, it will decrease very rapidly in the expansion and will become equal to the more slowly decreasing e (actual vapour pressure of the air). At that point the air has become saturated. The process that has occurred is indicated in the vapour pressure diagram of Figure 3 by the dashed line from P (initial state of air) to Q (saturation). This occurs in the atmosphere at a given level, depending on the initial humidity and temperature of the air; this is the level of the cloud base. From there on, the air may continue ascending, but then condensation is taking place. On the diagram, the representative point slides along the saturation curve from Q to R. We can no longer apply (23), but must derive the corresponding equation for a cloud.

Proceeding as before, we start now from Equation (13) and the adiabatic condition $\delta Q = 0$. We then obtain

$$C_p \, dT - V \, dp + L_v \, dN_v = 0 \tag{26}$$

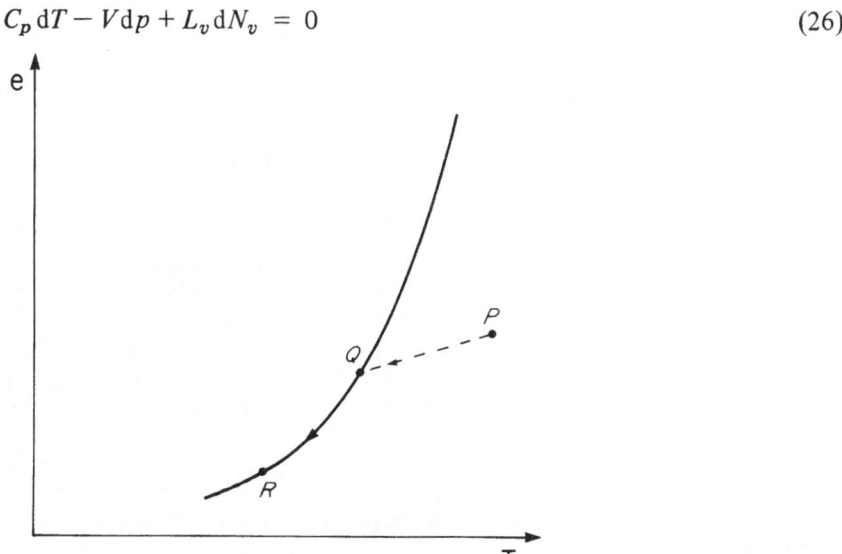

Fig. IV-3. *Cooling and condensation by adiabatic expansion. T =* temperature; *e =* water vapour pressure; *P =* initial state of air; *Q =* saturation; *R =* final state of air. Along *QR, e* decreases partly because of condensation and partly due to the expansion.

Dividing by T and introducing the gas law, this takes the form

$$C_p \, \mathrm{d} \ln T - R \, \mathrm{d} \ln p + \frac{L_v}{T} \, \mathrm{d} N_v = 0 \tag{27}$$

Here $\mathrm{d} N_v$ can be replaced by (17), which leads to the formula

$$(C_p + L_v^2 e / R p T^2) \, \mathrm{d} \ln T - (R + L_v e / p T) \, \mathrm{d} \ln p = 0 \tag{28}$$

which relates the variation of T with that of p. Comparing (27) with (22), we see that the difference lies in the term $(L_v/T) \, \mathrm{d} N_v$. This indicates the influence of the released latent heat, which will partially compensate the decrease in temperature. The result is a curve of $T = f(z)$ (z = altitude) less steep for the saturated case; the higher the humidity, the higher the condensation, and the less steep the curve (i.e. the less sharply the temperature decreases). Air that rises first unsaturated, then saturates and continues rising, will follow a curve such as that indicated in Figure 4.

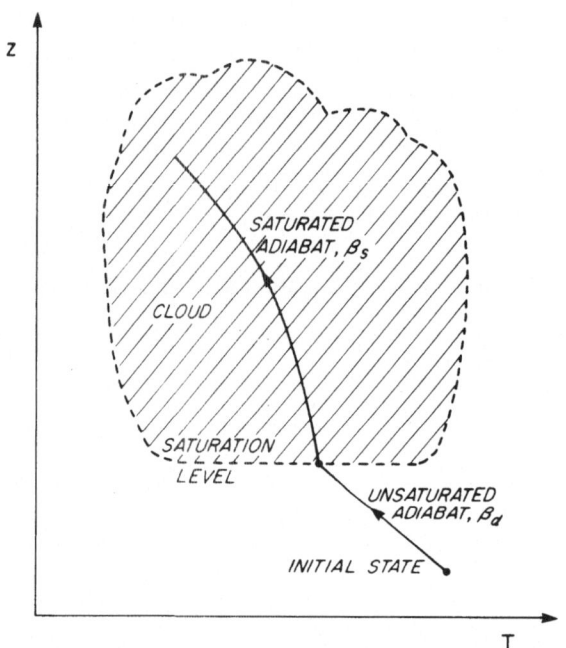

Fig. IV-4. *Ascent of air in the atmosphere.* T = temperature; z = height. Neglecting mixing with the environment, ascent implies an adiabatic expansion. While the air is unsaturated, the temperature rate of decrease is given by $\beta_d = 9.8 \, \mathrm{K/km}$. When the air becomes saturated, a cloud starts forming and the temperature continues dropping at a smaller lapse rate β_s.

The rate at which the temperature decreases during an adiabatic ascent, for dry air, is 9.8 K/km, and only slightly less (up to a few %) for moist, unsaturated air. For saturated ascent, this rate varies with T and p, but is always less than the previous value, and can be as low as 3 K/km. We shall call $\beta = - \mathrm{d} T / \mathrm{d} z$ the temperature *lapse rate*. For dry air, and approximately any unsaturated air, $\beta_d = 9.8 \, \mathrm{K/km}$; for saturated expansion, $\beta = \beta_s$.

7. Horizontal Mixing

If two different adjacent air masses mix, the process occurs essentially at constant pressure. If no condensation takes place, the result is air with a temperature and a molar ratio N_v which are the weighted averages (weighted over the number of moles) of the values for the two air masses.

It may happen that two unsaturated air masses produce condensation when mixing. This occurs if one of the masses is hot and humid, while the other is cold, and it is due to the curvature of the vapour pressure curve. Figure 5 illustrates the process. It can be shown that the representative point of the mixture lies on the straight line joining the two mixing air masses. The figure shows that this may lead to supersaturation, with the subsequent condensation. This can produce a fog, called *mixing fog*. Jet aircraft trails are a case of mixing fog, where the hot and humid gases from the motor exhausts mix with the outer cold (and therefore dry) air.

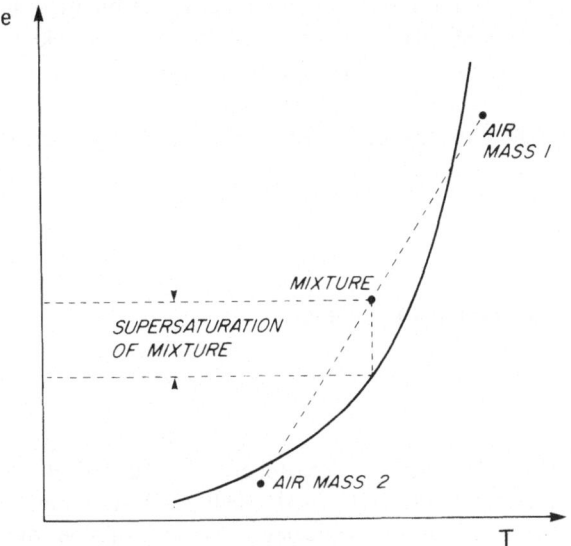

Fig. IV-5. *Mixing fog.* T = temperature; e = water vapour pressure. Due to the curvature of the saturation vapour pressure curve, mixing can lead to supersaturation.

8. Vertical Mixing

When a region of the atmosphere is unstable, convection may ensue, with vertical ascents and compensating descents of air. This may lead to a thorough mixing of a whole layer, particularly in areas with strong insolation.

Let us assume that such a thorough mixing has taken place through a layer of a given thickness, say between p_1 and p_2. We want to find out what will be the final vertical distribution of temperature. We start by considering an infinitesimal layer of thickness dz at the pressure p, and we imagine that we bring it to the level of 1000 mb, adiabatically. Its temperature will then become equal to its potential temperature θ, by definition of this parameter (cf. Equation (24)). Now let us imagine that we bring the whole finite layer to the 1000 mb level, infinitesimal layer by infinitesimal layer, and that we mix it thoroughly at that level. According to §7, the result will be that all this air will acquire the temperature $\bar{\theta}$ equal to the mass-averaged potential temperature. Now we imagine that we redistribute all the air, infinitesimal layer by infinitesimal layer, to their original situations in the interval of pressures p_1 to p_2*; in so doing, every parcel of air conserves the value of the potential temperature, now equal to $\bar{\theta}$ for all the air. Thus the final result is a finite layer (between p_1 and p_2) with constant potential temperature $\bar{\theta}$ throughout the whole thickness. This corresponds to a distribution of T according to (24), i.e. the final vertical variation of the temperature will be equal to the temperature variation that a parcel of air with potential temperature $\bar{\theta}$ undergoes if raised through that height, and we saw that this is given by a lapse rate $\beta_d = 9.8$ K/km. This is expressed by saying that the result of vertical mixing without condensation is a 'dry adiabatic layer'.

Notice that in this result we are considering the instantaneous final stratification of the layer, rather than the process undergone by rising air. But both will be represented by the same $T = f(z)$ curve, and this will be the lower part of the curve in Figure 4, labelled as 'unsaturated adiabat'. ('Unsaturated adiabat' and 'dry adiabat' are to be taken as equivalent names for the same curve, since we neglect the influence of humidity when the air is under saturation).

If the mixing is accompanied by condensation, starting at a certain level, the result will be described, statically, by the same total curve in Figure 4 that represented the ascent process.

9. Vertical Stability. Conditional and Latent Instability

If we have a mass of homogeneous fluid, at uniform T, in a container of laboratory dimensions, and we heat a portion of the lower layer, we create a vertical instability. The heated fluid becomes less dense than its surroundings, thereby experiencing an upwards force (buoyancy), and starts rising. Since in the atmosphere we deal with dimensions of the order of km or tens of km, both T and p vary with height, and therefore the density ρ also varies. We want to find now a criterion to decide if the atmosphere, or a portion thereof, is in equilibrium with respect to vertical displacements.

In order to find that criterion, we can apply the method of assuming *virtual displacements* cf a parcel (any portion of the atmosphere, at a given level) and studying whether the consequences of such displacements will be a restoring force tending to bring back the parcel to its original position or a force tending to increase further the displacement. In the first case, we have an equilibrium condition, and the atmosphere is stable; in the second case, the atmosphere is unstable.

In doing this, a number of simplifying assumptions are implicitly made: (1) that the parcel does not mix with the surroundings, (2) that its motion does not disturb the environment, (3) that the process is adiabatic and (4) that at every instant the pressures of the parcel and of the environment are equal for a given level.

---

* Notice that each infinitesimal layer will go to its original pressure p, but not in general to the same original height.

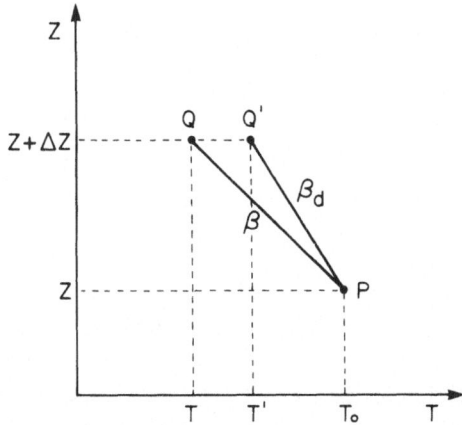

Fig. IV-6. *Vertical instability*. If the geometric temperature lapse rate β at height z is larger than β_d, an unsaturated parcel rising Δz from P will have at Q' a temperature $T' > T$, where T is the temperature of the surrounding air (Q); it will tend to continue rising.

Now let us consider a level in a non-saturated atmosphere at which β is the *geometric* lapse rate. By this we mean the vertical temperature stratification, not to be confused with the *process* lapse rate that gives the change of T with z when a parcel rises or descends. Let us assume that $\beta > \beta_d$. If we imagine a virtual displacement, we shall have the situation of Figure 6. Here P is the representative point in a graph z, T of the atmosphere at level z. At a level $z + \Delta z$ (Δz can be considered here as infinitesimal), the representative point is Q, and the temperature is $T_0 - \beta \Delta z$. A parcel displaced in Δz from P, on the other hand, will follow an adiabat β_d, and at $z + \Delta z$ will have a temperature $T_0 - \beta_d \Delta z > T_0 - \beta \Delta z$. Because the pressure will be equal for parcel and environment, the density ρ' of the parcel will be smaller than the density ρ of the environment, and the parcel will experience a positive buoyancy, i.e. an upward force. Because this buoyancy is positive, after the virtual displacement the parcel will tend to continue rising. It is clearly an unstable situation.

If $\beta < \beta_d$, the case is that of Figure 7. Then the buoyancy is negative, so that the displaced parcel will tend to go back to the original level z. It is a stable case.

Notice that with $\Delta z < 0$ the same conclusions would be reached. If the atmosphere is saturated, the same type of argument can be made, but using β_s (rather than β_d) for the process lapse rate of the parcel.

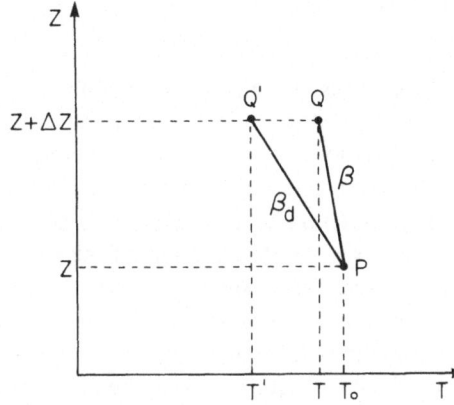

Fig. IV-7. *Vertical stability*. If the geometric temperature lapse rate β at height z is smaller than β_d, an unsaturated parcel rising Δz from P will have at Q' a temperature $T' < T$, where T is the temperature of the surrounding air (Q); it will tend to sink back to z.

Thus we have the following conditions:

$$\left.\begin{array}{c} u \\ \text{Unsaturated } \beta \lesseqgtr \beta_d \\ s \\[1em] u \\ \text{Saturated } \beta \lesseqgtr \beta_s \\ s \end{array}\right\} \tag{29}$$

where the upper inequality signs correspond to the unstable case (u) and the lower one to stability (s). If the equality sign holds, the equilibrium is indifferent: the displaced parcel remains where it is left. All these cases are summarized schematically in Figure 8. If the lapse rate of the atmosphere at P (i.e. at level z and pressure p) lies to the left of β_d, the atmosphere is *absolutely unstable* (at that level); if it lies to the right of β_s, it is *absolutely stable*. If it lies between β_d and β_s, it is said to be *conditionally unstable*, meaning that it is stable if not saturated, but unstable if saturated.

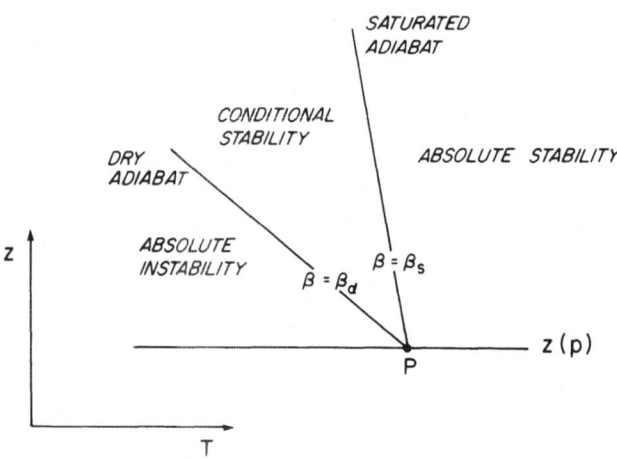

Fig. IV-8. *Conditions of vertical stability*. The diagram summarizes the stability conditions for both unsaturated and saturated air (see text). T = temperature; z = height.

So far, we have considered only infinitesimal displacements. Air can be forced to rise even in a stable atmosphere (for instance, if wind blows against a mountain slope) and it may eventually follow a process curve such as is indicated in Figure 9. Here the curve β indicates the geometric stratification of the atmosphere. The parcel considered follows β_d until saturation at Q (*lifting condensation level*) and then follows the saturated adiabat β_s. At R (*level of free convection*), the representative point of the parcel crosses to the right of the curve β, and the buoyancy becomes positive. From R on, the parcel is accelerated upwards. This region, indicated in the figure by the shaded area, is the region of *latent instability*. In this type of instability, the parcel may actually become mixed to a large extent with the environment, so that the conclusions can only be qualitative.

The study of the vertical stability is very important in weather forecasting. Instability

Fig. IV-9. *Latent instability*. $T =$ temperature; $z =$ height. β indicates the temperature stratification of the atmosphere. A parcel rising from P will follow an unsaturated adiabat (lapse rate β_d) to Q (*lifting condensation level*) and then a saturated adiabat (lapse rate β_s). The parcel must be forced to rise up to R (*level of free convection*); from there on, it is accelerated by a positive buoyancy. The shaded area indicates the *latent instability*.

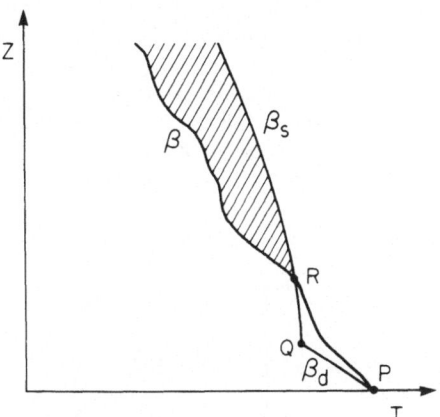

is associated with weather perturbations. Thus, if conditions are such that strong vertical motions occur, 'convective' clouds are formed: first *cumuli*, and when these develop further, *cumulonimbi*, i.e. huge clouds with thickness of the order of 10 km, whose tops become glaciated when their temperatures fall below $-40°$C. Showers and hail develop in these clouds, and the strong vertical air currents in them can reach velocities of several tens of meters per second.

10. Potential or Convective Instability

In certain cases (for instance, at the approach of a cold front), whole layers of atmosphere can experience a lifting. It is important to consider how this will affect the stability conditions.

Let us consider first the case when saturation does not occur. AB is a layer of geometric lapse rate β (Figure 10). As there is no saturation, lifting will bring A to A' and B to B' along process dry adiabats β_d. The thickness of $A'B'$ (i.e. of the layer after lifting) will be larger than that of AB, because the pressure decreases with height and therefore the layer expands. The geometric lapse rate has become β', and it is clear that, while remaining $< \beta_d$, it will be $\beta' > \beta$. The layer has become less stable. If at the initial level air is converging laterally towards the zone of lifting, $\Delta z'$ will become still larger as compared with Δz, simply because of mass conservation (the layer stretches vertically to allow space for the converging air); the result will be a still larger instabilization.

Therefore: *both ascent of a layer and horizontal convergence decrease the stability*. Reciprocally, descent (or subsidence) and divergence, increase the stability.

Now let us consider layer lifting when saturation occurs. Two cases can happen. In the first one, humidity decreases with height within the layer, so that the lower part of the layer becomes saturated first. If the layer is AB (Figure 11), both A and B will follow first

91

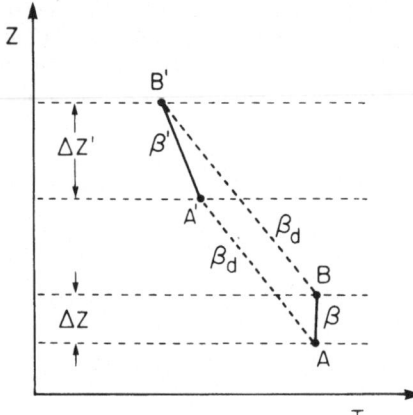

Fig. IV-10. *Ascent of a layer*. T = temperature; z = height. AB and $A'B'$ indicate the (geometric) lapse rate within a layer before and after rising. Due to the decrease in pressure, the thickness increases ($\Delta z' > \Delta z$) and β tends towards the value β_d. Horizontal convergence will make this effect even more pronounced.

Fig. IV-11. *Potential instability*. T = temperature; z = height. Humidity decreases with height within the layer AB. If the layer rises, its lower portions become saturated earlier and this leads to an unstable lapse rate β' as indicated by $A'B'$.

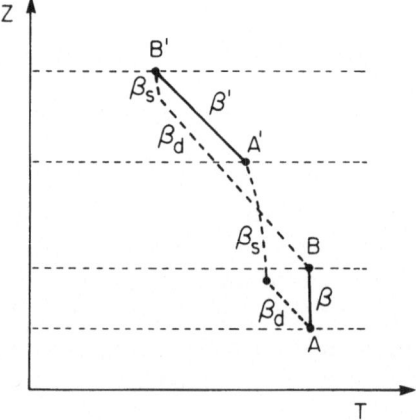

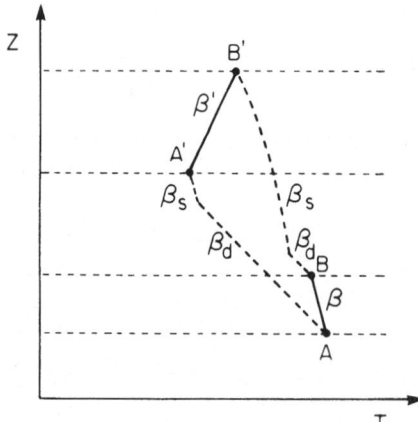

Fig. IV-12. *Potential stability*. T = temperature; z = height. Humidity increases with height within the layer AB. If the layer rises, its upper portions become saturated earlier and this leads to more stable lapse rates β', as indicated by $A'B'$.

a dry adiabat β_d, and then a saturated adiabat β_s, to the final points $A'B'$, but a parcel starting at A will saturate earlier than a parcel starting at B. It is clear that the layer not only loses stability, but can become absolutely unstable. This is called *potential* or *convective instability*.

If the opposite happens, i.e. that the humidity increases with height, the layer will become more stable with lifting. It is the case of *potential* or *convective stability*. This is exemplified in Figure 12.

92

Chapter IV: Questions

Q1. Are large supersaturations of water vapour observed in the atmosphere? Explain.

Q2. Why do dew and frost form during the night?

Q3. Can the dew point be observed below $0°C$? The frost point above $0°C$?

Q4. If the total pressure of moist air changes, how does its partial pressure of water vapour change, assuming that there is no mixture with the environment?

Q5. Can a fog result from mixing two unsaturated air masses? Explain.

Q6. Does the potential temperature change when air rises without condensation in the atmosphere? When it rises with condensation? When it is cooled over the ground?

Q7. Why does humid air form clouds when rising?

Q8. Why are cloud bases horizontal, while the tops may be rounded?

Q9. What happens if cloudy air sinks?

Q10. What is the result of vertical mixing of an unsaturated atmospheric layer, as regards temperature stratification?

Q11. During the afternoon on hot summer days, the temperature lapse rate is often that of a dry adiabat in the lowest layers over the ground. Explain.

Q12. A saturated air parcel is rising within a cloud. When will it stop rising?

Q13. Do temperature inversions indicate stable or unstable layers in the atmosphere?

Q14. Smoke from stacks disperses sometimes in long uniform horizontal plumes, while at other times it mixes rapidly with air in almost vertical disorganized puffs. Explain.

Chapter IV: Problems

(Any necessary constants not given in the statement of a problem will be found in the Table of Constants on pages x–xi)

P1. Using Figure II, 1, calculate the molar fraction of water vapour in air saturated at 20°C and 1000 mb.

P2. Use the Clausius–Clapeyron equation to find the dew point and the frost point of air with 2.0 mb of water vapour pressure.

(*Note*: The curve of saturation vapour pressures of ice is described by a similar equation to that for water, substituting L_s, the molar heat of sublimation, for the molar heat of vaporization L_v).

P3. Air is losing heat by conduction to a colder ground and a fog is developing. If at the time when the air is at 0°C the temperature drops at the rate of 0.6 K/hour, what is the rate at which heat is being lost, per unit mass of air? The ground pressure is 1000 mb.

P4. During the night, the air above the ground loses heat by radiation and conduction to the ground, at the rate of 10 cal/kg.min. The initial temperature and water vapour pressure are 10°C and 10 mb.

(*a*) What is the rate of cooling (in K/min) before saturation?

(*b*) At what temperature does the air become saturated?

(*c*) What is the initial rate of cooling, once the fog starts forming?

Use Figure II, 1 or tables to find saturation vapour pressures.

P5. Air at 20°C containing 20 mb of water vapour pressure is cooled both by conduction to the ground and by long wave radiation, until it reaches 10°C. Has a fog been produced, and if so, what is its content in liquid water, given in grams per cubic metre of air? The total pressure is constant and equal to 1000 mb.

P6. A cloud has a temperature of 0°C at 900 mb of pressure. It absorbs heat, with the result that the temperature increases to a final value of 10°C. In spite of the water evaporated (which has maintained the air saturated), there is still liquid water remaining at the end, in the form of droplets. The pressure remains constant. How much did the enthalpy of one kilogram of cloudy air change? Saturation vapour pressure at 0°C: 6.1 mb; at 10°C: 12.27 mb.

P7. Air rising from the ground forms a cumulus cloud. Condensation starts (cloud base) when its temperature reaches 10°C. Assuming that there was no mixing with the surroundings, what was the relative humidity of the air at the ground, where the temperature is 25°C? The relative humidity is equal to the vapour pressure divided by the saturation vapour pressure. Use Figure II, 1 for values of saturation vapour pressures.

P8. A mass of air rises from close to the ground, at 1000 mb pressure, to the 500 mb level. There is no condensation. Assume that the air does not mix and does not exchange heat with the surroundings. Its initial temperature is 20°C.
(a) What will be the final temperature of the air mass?
(b) What are its initial and final potential temperatures?
(c) Assuming that the temperature lapse rate of the atmosphere (surroundings) is $\beta = 6\,K/km$, constant in height, calculate the altitude of the 500 mb level, and
(d) the temperature (of the surroundings) at that level.

P9. An ascending mass of air becomes saturated at 900 mb and 5°C. About 200 m above the base of the cloud, the pressure and temperature decrease to 880 mb and 4°C. Considering that the differences can be treated approximately as differentials, calculate (a) the decrease in molar fraction of water vapour, and (b) the liquid water content of the cloud at 880 mb, in grams of water per cubic metre of air.
Values of water vapour saturation pressures can be obtained from Figure II, 1 or from tables.

P10. Two equal air masses mix at constant pressure. Their temperatures and water vapour pressures are:

$$T_1 = 23.8°C; \quad e_1 = 25.5\,mb$$

$$T_2 = -6.4°C; \quad e_2 = 2.1\,mb$$

Use Figure II, 1 to
(a) determine the relative humidities of both air masses (relative humidity = ratio of water vapour pressure over saturation pressure).
(b) determine whether this mixture results in a fog. The representative point of the mixture, if there was no condensation, would be the middle point of the straight line joining those representing the initial air masses. If this results above saturation, a fog will form. Can you, in this case, take the difference between the final e and the saturation vapour pressure e_s at the final temperature, to calculate directly the amount of liquid water in the fog? Explain.

P11. Show that if a layer of atmosphere comprised between p_1 and p_2 is thoroughly mixed (e.g., as a consequence of convective activity), the uniform potential temperature in the final state is given by

$$\bar{\theta} = \frac{\int_{p_2}^{p_1} \theta\,dp}{p_1 - p_2}$$

where $\theta = \theta(p)$ gives the value of the potential temperature at each level, before mixing.

P12. Assume that the air close to the ground is at a pressure $p_0 = 1000$ mb and a temperature T_0. The temperature decreases with height at a constant lapse rate, so that at 500 m it reaches T. Consider the two following cases:

$$(i)\ T_0 = 30°C, \quad T = 24°C$$

(*ii*) $T_0 = 20°C$, $T = 17°C$

(*a*) Find the pressure p at 500 m in each case.

(*b*) Compute the potential temperature θ at the ground and at 500 m in each case.

(*c*) Compute the lapse rate β in each case; explain whether the air is stable or not, and in the latter case, what will happen.

(*d*) State the stability conditions in terms of the variation of potential temperature, rather than in terms of β.

V. Cloud Physics

1. Introduction. Classification of Clouds

In the previous chapter we have considered the thermodynamic aspect of water condensation due to the ascent of air, and how this is related to the vertical stability in the troposphere. The visible result of that ascent and condensation is of course the formation of clouds; this is tantamount to saying that clouds often constitute visible indicators of vertical motions in the atmosphere. Thus the description and classification of clouds provide something more than simply a systematic account of shapes and characteristics; it helps, in conjunction with other information, to identify the *synoptic* situation, i.e. the state of a large portion of atmosphere at a given time, and thereby assists the meteorologist in weather forecasting. For instance, the slow development of extensive layer clouds are indicative of a slow massive ascent of atmosphere, isolated clouds of large vertical development point to conditions of vertical instability, the approach of a cold front (cf. Ch.VII) is heralded by a whole succession of different types of clouds, etc.

Therefore we shall make a brief reference to the main types of clouds, and only then shall we go into the principal object of this chapter, which is the microphysical consideration of how clouds and precipitation develop (§2 and following).

The classification has been summarized in Table 1, which we shall now comment upon. In the adopted names, several words, prefixes or suffixes appear repeatedly. *Cirrus*, *cirro-* indicate a cloud consisting of ice crystals, and consequently appearing only at high levels where the temperature is low enough. *Strato-*, *(-)stratus* mean a layer cloud, whether continuous or showing a certain structure. *Alto-* refers to middle height clouds. *Cumulo-*, *(-)cumulus* indicate vertical circulation with localized updraughts, whether isolated or repeating in a pattern. *Nimbo-*, *(-)nimbus*, indicate a type of cloud producing precipitation. Cumulus and Cumulonimbus differ from the other clouds in that they show considerable vertical development over a limited area; they constitute the category of convective clouds, as opposed to *layer* clouds. Other characteristics and descriptions can be read from Table 1. The nomenclature is international, and the World Meteorological Organization has published an International Cloud Atlas with typical photographs for each kind of cloud. Some examples of different types of clouds are given in Figures 1 to 6.

2. Condensation of Water Vapour in the Atmosphere. Cloud Condensation Nuclei (CCN)

If atmospheric air is purified from all its aerosol content, for instance by filtration, condensation of its water vapour occurs with difficulty. Thus, if it is cooled to its dew point, i.e. to saturation, condensation does not start. It has to be cooled much further, until the ratio of its water vapour pressure to the saturation vapour pressure ('saturation ratio') reaches a value around 4, before condensation starts taking place. At this value, liquid

TABLE 1. Cloud classification.

Cloud (Genus)	Height range*	Composition	Description
Cirrus	High	Ice	White bands or delicate filaments or patches with fibrous or silky sheen appearance, or both.
Cirrocumulus	High	Ice	White patch or layer of more or less regularly arranged small elements in the form of grains, ripples, etc.
Cirrostratus	High	Ice	Whitish veil with little structural details, covering smoothly a large extension of the sky.
Altocumulus	Middle	Water	White or grey layer of finely regularly-arranged small cloud elements, patch of small elements, smooth patch with well defined outlines, or layer of cumuliform tufts.
Altostratus	Middle	Water	Greyish or blueish layer of fibrous or uniform appearance, covering a large extension of sky.
Nimbostratus	Low	Water	Grey thick layer, often dark, in general with falling rain or snow.
Stratocumulus	Low	Water	Grey or whitish layer with dark elements, often regularly arranged.
Stratus	Low	Water	Grey layer with fairly uniform base, sometimes precipitating drizzle or snow grains.
Cumulus	Usually low base. Vertical development of several km.	Water	Detached clouds, dense, with sharp outlines, developing vertically ('cauliflower' clouds). Precipitation may develop, of the showery type.
Cumulonimbus	Usually low base. 5 to 12 km vertical development; may reach tropopause.	Water in the main body; ice in top part	Heavy and dense cloud of large vertical extent. The upper part is glaciated, showing a fibrous appearance and generally extending horizontally in an *anvil* shape or in a plume. Heavy precipitation develops – rain or hail – and it sometimes becomes electrified, producing lightning ('thunderstorms').

* In middle latitudes, the meaning of these ranges is as follows: high = 7 km to tropopause; middle = 2 to 7 km; low = surface to 2 km.

98

Fig. V-1. *Cirrostratus.* The fibrous appearance is typical of the high ice clouds.

Fig. V-2. *Altocumulus and stratocumulus.* Clouds appear in two layers in this photograph. The upper one of middle clouds – altocumuli – and the lower one forming a deck of stratocumuli (low clouds).

Fig. V-3. *Cumulus and stratocumulus*. **Typical low** clouds of small vertical development. **The isolated** elements are small cumuli. **Towards the bottom right**, they become assembled into a layer of strato-cumuli.

Fig. V-4. *Cumulus*. Successive air masses rise **through the cloud, emerge at the top as rounded towers** until they lose buoyancy, mix with surrounding air and decay. The cloud grows by this process to a height that depends on the stability conditions of the atmosphere (cf. Ch.IV).

Fig. V-5. *Cumulonimbus*. Among smaller cumuli, a cumulonimbus rises and attains much higher levels. The well defined, rounded towers at the left indicate rising air masses. At those heights, the cloud becomes glaciated and spreads over large distances (anvil); this occurs in this example towards the right, because of the wind shear (derivative of wind velocity with respect to height) at those levels.

water starts nucleating on the negative ions, always present. At still higher saturation ratios, around 6, it also nucleates on positive ions. But, as we shall see, these supersaturations are never reached in the atmosphere.

The particles of the atmospheric aerosol act as condensation nuclei (CN) at much lower supersaturations. In fact, there is a whole spectrum of activity, the larger particles acting as CN at lower supersaturations than the smaller ones. Thus, when air rises, cooling by adiabatic expansion, it reaches saturation and, as soon as the water vapour exceeds the saturation value, water starts condensing on the largest hygroscopic nuclei. As supersaturation increases, more and more nuclei become active, until the effect of condensation in depleting the air of water vapour becomes larger than the effect of adiabatic cooling, so that a maximum is reached in the supersaturation. From there on, supersaturation decreases and no more nuclei are activated. The fraction of activated nuclei is small: of the order of 100 (or at the most a few hundreds) cm^{-3}, while the total nuclei content may be one or several orders of magnitude larger. The number of activated nuclei will correspond to the number of cloud droplets formed. The CN that become active in the formation of clouds are called *cloud condensation nuclei* (CCN).

We shall now consider this process with some more detail.

The saturation value of the water vapour pressure e_s is defined as the pressure at which the vapour is in equilibrium with a *plane* surface of water, although this condition of absence of curvature is usually left implicit. Surfaces of very high curvature, i.e. very small radius of curvature, as in small droplets, have higher values of the saturation vapour pressure. This is connected with the surface energy of the droplets. It can be derived from a thermodynamic argument that this pressure e_r can be expressed by Lord Kelvin's formula:

$$e_r = e_s \exp\left(\frac{2M\sigma}{\rho RTr}\right) = e_s \exp\left(\frac{\text{const}}{r}\right) \tag{1}$$

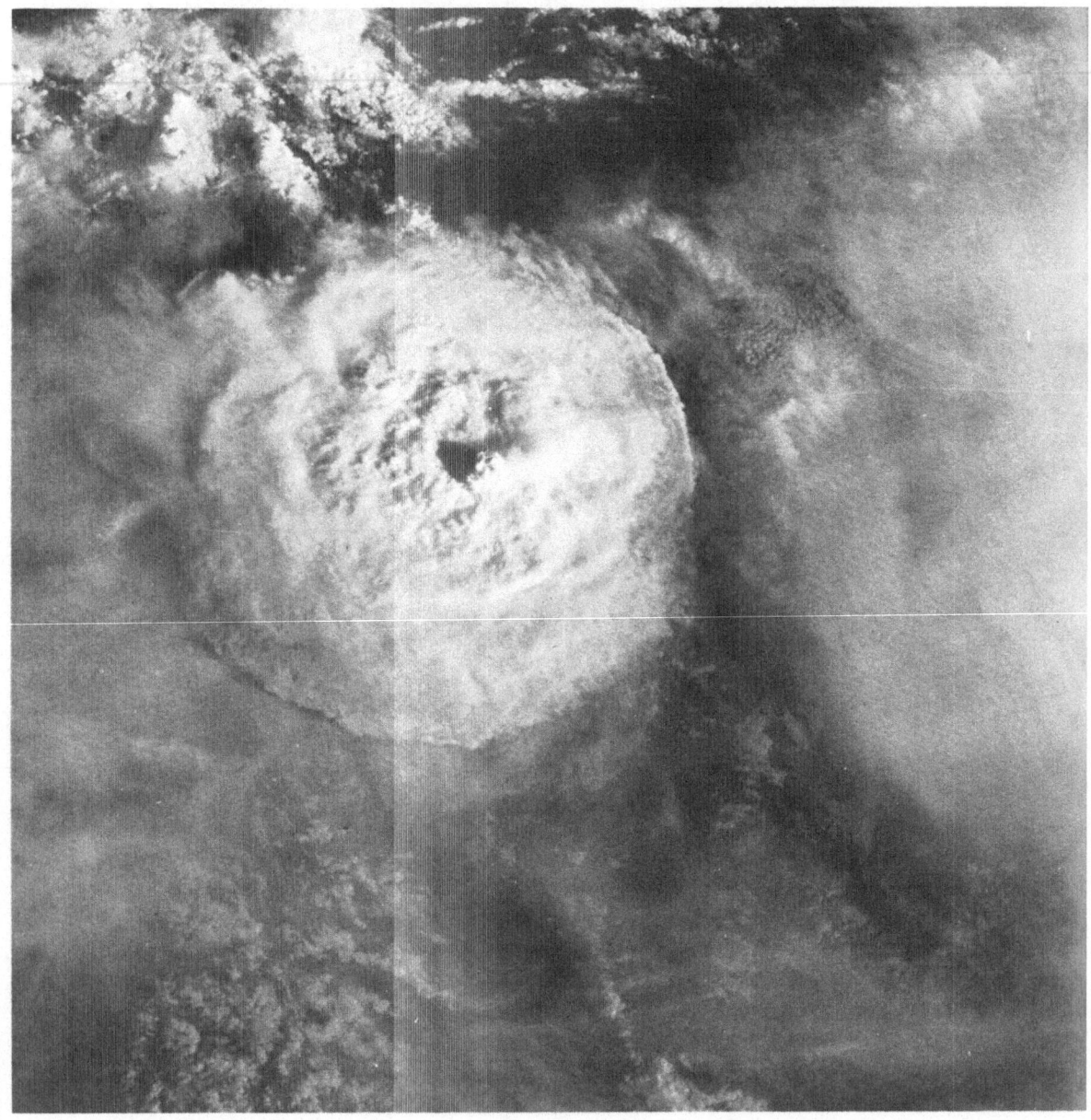

Fig. V-6. *Cumulonimbus.* Near vertical view over the top of a thundercloud over South America, from the Apollo 9 spacecraft. The round shape corresponds to the anvil. Burgeoning towers can be seen rising at its centre, the main one high enough to cast a large shadow. Cumuli can be seen at the top left corner; cirrus clouds (mainly cirrostratus) cover most of the rest of the picture, around the cumulonimbus.

where M = molecular weight = 18.02 g/mol for water; ρ = density = 10^3 kg/m^3; R = gas constant; T = absolute temperature; σ = surface tension = 0.072 N/m for water; r = drop radius.

The values corresponding to (1) are plotted in the upper curve of Figure 7 (for $0°C$) in terms of Supersaturation vs. r, where Supersaturation $= (e_r/e_s) - 1 = S - 1$ ($S =$ saturation ratio).

If the droplet has been formed on a hygroscopic nucleus, for instance of NaCl, a second effect has to be considered: that of the non-volatile solute on the vapour pressure. This effect is given, for dilute solutions, by Raoult's law:

$$e'_s = e_s N_w \tag{2}$$

where $e'_s =$ water vapour pressure of a solution (plane surface); $N_w =$ molar fraction of water in the solution.

This is a reduction of the vapour pressure, because $N_w < 1$. For a constant mass of solute, smaller and smaller radii of the drop mean higher and higher concentrations

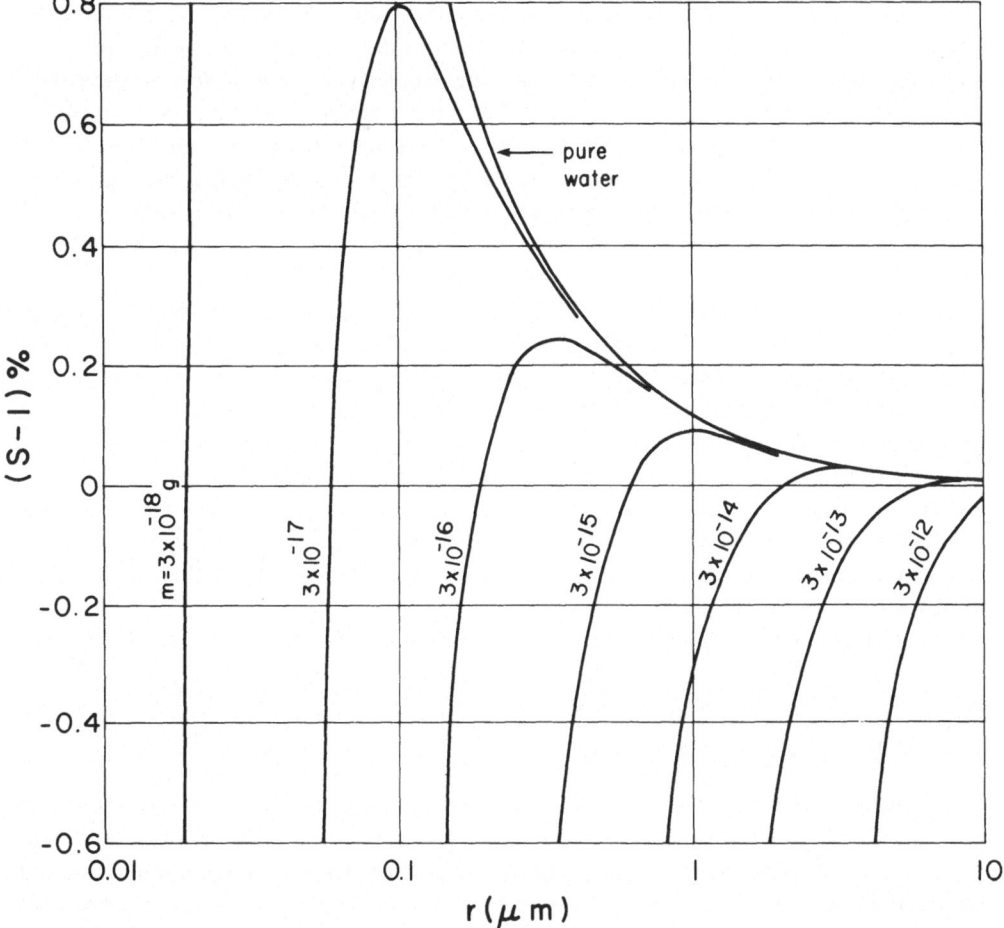

Fig. V-7. *Activation of condensation nuclei.* $(S - 1) =$ supersaturation; $r =$ drop radius. The upper curve gives the supersaturation corresponding to the vapour pressure of pure water droplets. The other curves correspond to droplets containing a fixed mass of sodium chloride, as indicated on each curve. The line for 0 supersaturation corresponds to the equilibrium vapour pressure over a plane surface of pure water. The curves are calculated for $0°C$.

(increasing as $1/r^3$); therefore, stronger effect. Raoult's law (2) ceases to be valid, but the effect is always in the same sense.

The effect of the solute, smaller than that of curvature at larger radii, becomes predominant for smaller radii. The result is a curve that, as we go from large to smaller radii, starts separating from the value given by (1) (upper curve), goes through a maximum, and then decreases sharply. This is represented by a set of curves in Figure 7, where the solute is assumed to be NaCl and the variable parameter given for each curve is the mass (in g) of NaCl dissolved in the droplet. The (spherical equivalent) radii of the CCN corresponding to the curves in the figure vary between $0.006 \mu m$ (for the smallest mass) to $0.64 \mu m$ (for the largest mass).

The curves of the figure represent the values of the equilibrium vapour pressures (for varying size of the droplet, and each curve for a given mass of solute). Let us follow one of the curves, for instance that corresponding to 3×10^{-16} g. When the vapour pressure in the air is such that it is at 0.6% below the saturation value over plane water (i.e. supersaturation $= -0.6\%$), the droplet comes to equilibrium at a radius of $\sim 0.15 \mu m$. If we now increase the vapour pressure in the air, the droplet will grow, following the curve. When the supersaturation reaches about 0.24%, the droplet is at the maximum of the curve (with the critical radius). If we now remain in a supersaturation slightly above that maximum, the droplet will continue to grow indefinitely, because now as it grows its own equilibrium vapour pressure decreases; therefore the atmosphere is supersaturated for that drop and vapour will condense on it. We say that the original nucleus (CCN) on which that drop formed has become activated.

When air rises in the atmosphere, it will first reach the saturation level (cf. Ch.IV); now the supersaturation is exactly 0. Then, as it rises slightly more, values of supersaturations such as those of the upper part of the figure are attained. All droplets (viz. their initial nuclei) whose maxima are below the supersaturation become activated, i.e. grow beyond the maximum. Then condensation over all these droplets becomes active enough to reduce the supersaturation; no more droplets become activated. Those on very small nuclei did not have a chance of overcoming the maximum, and revert to their small equilibrium sizes corresponding to the left branches of the curves. These remain therefore as wet aerosol particles, but are not considered as cloud droplets (their sizes will be mostly under $0.1 \mu m$ radius).

Similar phenomena will occur when air cools below its dew point, and a fog is formed.

3. Growth of a Drop by Condensation

In the previous section we have considered how a cloud droplet starts. We are interested now in following its growth and in understanding the evolution that eventually results in the formation of a raindrop, i.e. in studying the development of precipitation. It should be realized that, between the initial CCN and the final raindrop, growth has occurred over many orders of magnitude in size. Figure 8 illustrates this point, giving typical values for sizes and concentrations.

The second stage in this evolution is the growth of a drop by condensation. We take our drop at the point where we left it in Figure 7, right after activation and with a radius of one or a few micrometers.

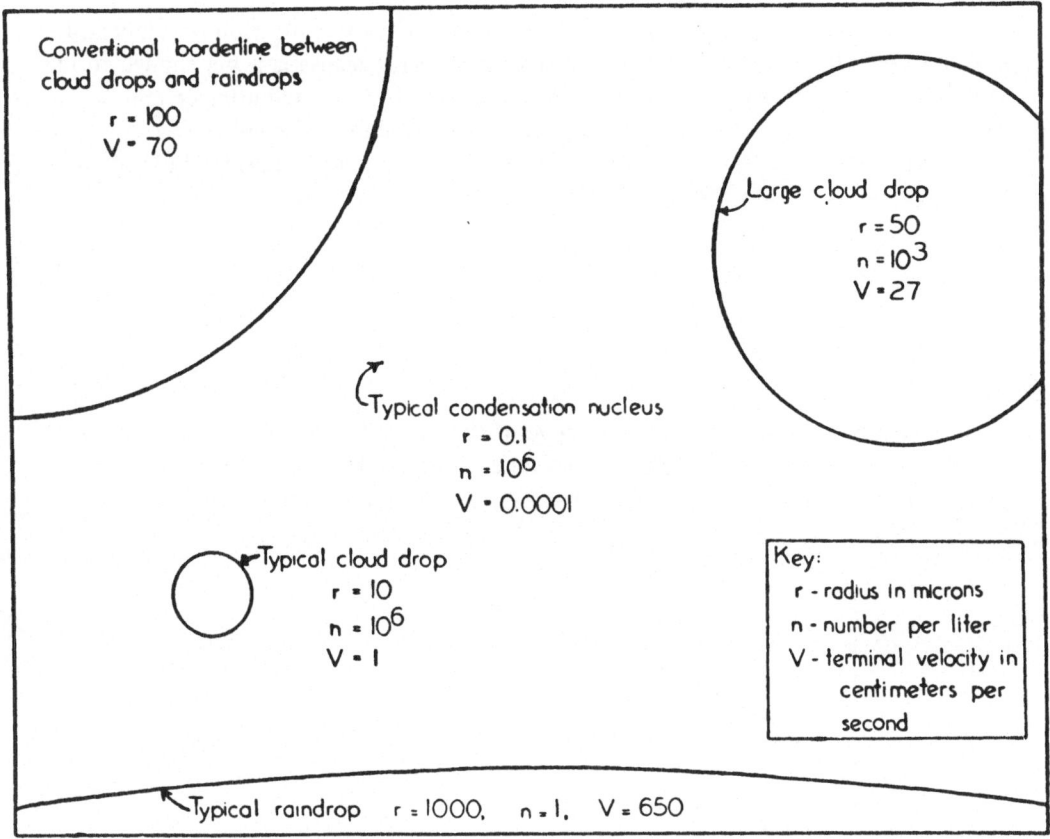

Fig. V-8. *Size range of elements involved in cloud physics.* Typical sizes, concentrations and falling velocities of nuclei, cloud droplets and raindrops.

The physical problem of growth by condensation can be considered as follows. Let us assume first that the drop is at rest, in an atmosphere whose water vapour pressure is larger than the equilibrium water vapour pressure of the drop. Then water vapour will diffuse towards the drop and condense on it. In so doing, it releases the latent heat of vaporization, causing the temperature of the drop to rise. This affects the equilibrium vapour pressure of the drop. After a certain time, a quasi-steady condition will be established, in which vapour diffuses towards the drop, and heat away from it, in a continuous way, and the relatively slow growth of the radius can be assumed not to disturb the boundary conditions.

The problem has spherical symmetry, and can be solved mathematically for both diffusions (of vapour and heat). We shall only see the result. If the drop has become large enough to neglect the effects of curvature and solute, the radius of the drop increases according to

$$r \frac{\mathrm{d}r}{\mathrm{d}t} = \frac{S-1}{f(T,p)} \tag{3}$$

105

where r = radius, t = time, S = saturation ratio and $f(T, p)$ is a function of temperature and pressure*. If the drop cannot be considered at rest, because it is big enough to fall with appreciable velocity in the air, the formula has to be corrected using certain 'ventilation factors'. This, however, does not change the situation in any essential way.

We notice that for steady conditions, where the whole right hand side of (3) could be considered as a constant C, we could write

$$r \frac{dr}{dt} = C \tag{4}$$

and integrate it to

$$r^2 = r_0^2 + 2Ct \tag{5}$$

where r_0 is the initial radius. This is a curve of the type of Figure 9 with a rapid initial growth (in terms of r), that soon becomes much slower. This is exemplified in Table 2, giving the times taken by droplets to grow in different conditions to several sizes. The life

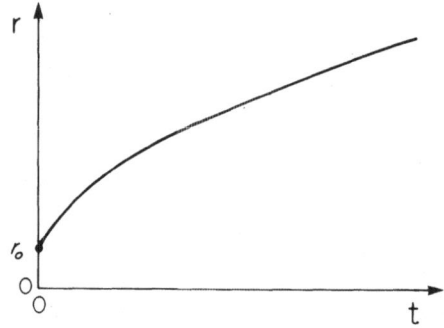

Fig. V-9. *Growth by condensation in a constant environment.* r = drop radius; r_0 = initial radius; t = time.

of many clouds can be as short as 10 to 20 minutes; even when it is much longer, still the life of each individual droplet will usually be limited to a period of the order of an hour, because it may be carried out of the cloud by the updraught or find its way to drier air by turbulent mixing. It becomes obvious that we cannot expect that condensation by diffusion alone will normally increase the size of cloud droplets beyond 15 to 20 μm in radius.

$$* f(T, p) = \frac{RT\rho}{DMe_s} + \frac{L_v \rho}{KTM} \left(\frac{L_v}{RT} - 1 \right)$$

where R = gas constant, T = absolute temperature, ρ = water vapour density, D = diffusion coefficient of water vapour in air, e_s = saturation vapour pressure, L_v = molar heat of vaporization, K = thermal conductivity of air, M = molecular weight of water. $f(T, p)$ varies rather slowly in the lower levels of the troposphere; for instance, its value is 1.0×10^6 s/cm² at 10°C and 700 mb, and 1.9×10^6 s/cm² at -12°C and 425 mb. As the temperature falls to low values in the higher troposphere, $f(T, p)$ increases more rapidly; for instance, at -40°C and 255 mb it amounts to 9.8×10^6 s/cm².

TABLE 2. Growth of drops by condensation.
Times calculated with formula (5) for the growth of a droplet from an initial radius of
1 μm to the indicated radius r. The two columns t_1 and t_2 correspond to $S - 1 = 0.1\%$
and $f(T, p) = 1.0 \times 10^{10}$ s/m² and to $S - 1 = 0.05\%$ and $f(T, p) = 1.9 \times 10^{10}$ s/m²,
representing rather favourable and rather unfavourable conditions, respectively.

r (μm)	t_1 (h : min : s)	t_2 (h : min : s)
2	0 : 00 : 15	0 : 00 : 57
5	0 : 02 : 00	0 : 07 : 36
10	0 : 08 : 15	0 : 31 : 21
15	0 : 18 : 40	1 : 10 : 56
20	0 : 33 : 15	2 : 06 : 21
30	1 : 14 : 55	4 : 44 : 42

4. Cloud Droplets

Until now we have only presented the basic ideas about how a cloud droplet is formed. In the atmosphere, the situation becomes complicated by many factors. For instance, the resulting population of droplets will depend on:
– the size distribution, concentration and chemical nature of the CCN;
– the updraught velocity of the air that will form the cloud. This can vary according to the type of situation between, say, 10 cm/s and 10 m/s;
– the mixing of cloudy air with its surroundings, by turbulent motion.

Observations have shown that the size spectra of cloud droplets are broader than expected from the theory. Interpretation of this fact has proved difficult, as it may depend mainly on the last factor mentioned above, which is not easily amenable to modelling and computation. Figure 10 presents some examples of size distributions observed for different types of clouds.

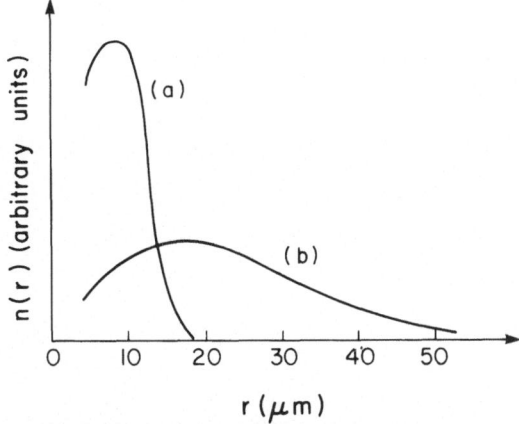

Fig. V-10. *Size distribution of cloud droplets.* The figure shows simplified examples of typical distribution curves $n(r)$ (distribution function over the radius, r) for (a) clouds with a narrow spectrum (e.g. fair weather cumuli, thin layers of stratocumulus or altostratus), and (b) clouds with a broad spectrum (e.g. large cumuli, cumulonimbus, nimbostratus). The total concentration of droplets is usually larger for (a) (several hundreds per cm³) than for (b) (less than one hundred per cm³).

107

5. Growth by Collision and Coalescence

If condensation is unable to increase the size of droplets beyond some 15–20 μm, how can these grow further, and explain the larger sizes seen in Figure 4 and the formation of raindrops?

It is obvious from Figure 10 that the spectra are far from monodisperse. They are broad enough to contain widely different radii. We have to consider now the falling velocities of these droplets.

If a drop starts to fall in the air, it will accelerate under the force of gravity. But at the same time, a drag force due to friction will tend to brake it. As this last force increases with velocity, soon the two forces come to an equilibrium, and from there on the velocity does not change any more. This is called the *terminal velocity*.

Terminal velocities are easy to compute for small sizes, when the motion is laminar (say below 20 μm radius); for larger sizes it becomes a difficult aerodynamical problem, further complicated in the raindrop size range by the deformations of the drops. However, experimental determinations have been made and terminal velocities are well known. Table 3 gives a few values over the whole range, and Figure 11 a complete plot in logarithmic scales.

TABLE 3
Terminal velocities of water drops in still air at 1 atm and 20° C.

Drop diameter (mm)	Terminal velocity (cm/s)
0.01	0.3
0.02	1.2
0.03	2.6
0.05	7.2
0.1	25.6
0.3	115
1	403
2	649
3	806
5	909
5.8	917

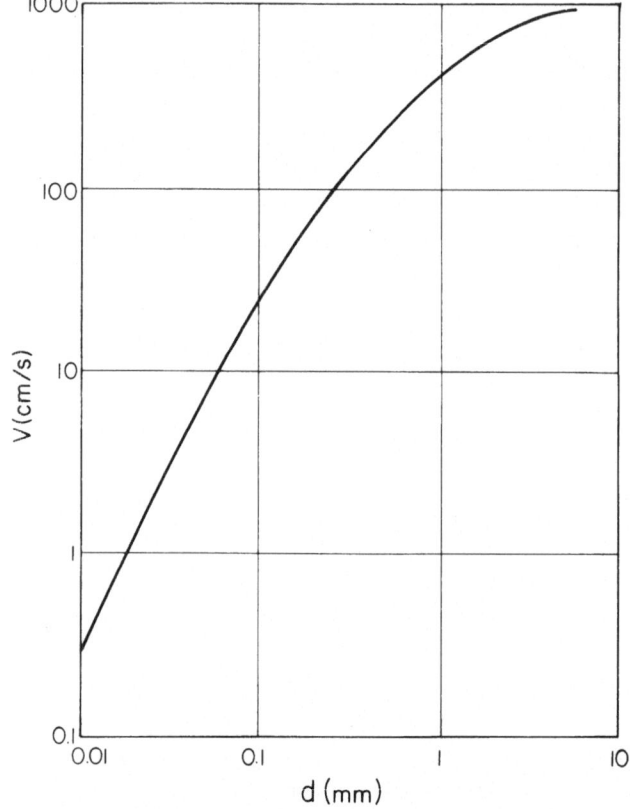

Fig. V-11. *Terminal velocities of water drops in still air at 1 atm and 20° C. V* = terminal velocity; *d* = diameter.

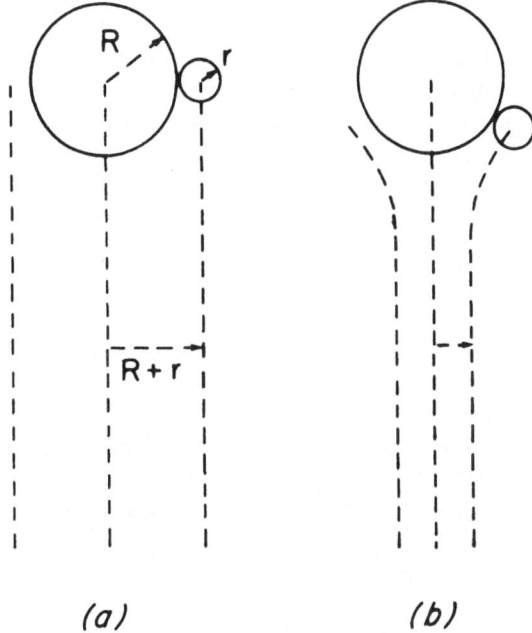

Fig. V-12. *Growth by coalescence and collection efficiency.* R = radius of larger drop; r = radius of smaller drop. (*a*) If the trajectories did not deviate from the vertical, any droplet with centre within the cylinder of radius $(R + r)$, whose axis passes through the centre of the bigger drop, would collide with the drop and eventually coalesce. (*b*) Aerodynamic effects change this situation as shown; here the smaller drop is displaced away from the axis, dragged by the air as it approaches the bigger one (which falls at higher terminal velocity). The collection efficiency E corrects for the change in cross section, as well as for non-coalescence collisions.

Let us go back now to our cloud droplets. It is clear that the larger droplets in the spectra of Figure 10 have appreciably higher terminal velocities than the smaller droplets. We have to consider the possibility that the former collide and coalesce with the latter, thereby increasing in size.

Let us assume, for simplicity, that one larger drop with terminal velocity V falls through a cloud of smaller uniform drops with terminal velocity v. It falls, therefore, at the relative velocity $(V - v)$ with respect to the small droplets. If these did not deviate at its passage, the larger drop would collide (and presumably coalesce) in one second with all the droplets contained in a cylinder of height $(V - v)$ and radius $(R + r)$, where R is the radius of the large drop and r that of the smaller one (see Figure 12a). Let us call w the water content of the cloud per unit volume (of the order of 1 g/m^3), i.e. the mass of all the droplets contained in unit volume. Then the mass added to the larger drop will be given by w times the volume of the cylinder mentioned above, which has the value $\pi(R + r)^2(V - v)$. But the problem is not so simple. For aerodynamic reasons, the droplets deviate at the passage of the larger drop, so that the relative trajectories actually look like Figure 12b. Therefore, we write empirically

$$\frac{dm}{dt} = E\pi(R + r)^2 w(V - v) \qquad (6)$$

109

where m = mass of the large drop and E is a correction factor called *collection efficiency*, which contains all the difficulties of the aerodynamic problem, and also corrects for collisions not leading to coalescence.

If m is put in terms of the radius R, Equation (6) becomes

$$\frac{dR}{dt} = \frac{Ew}{4\rho} \left(1 + \frac{r}{R}\right)^2 (V - v) \tag{7}$$

which for $R \gg r$ simplifies to

$$\frac{dR}{dt} = \frac{EwV}{4\rho} \tag{8}$$

where ρ is the density of water.

E depends on both R and r. The important result has been obtained that for R decreasing below $20 \, \mu m$, E decreases very sharply, making the collision mechanism for growth very inefficient. Above $20 \, \mu m$, E rapidly approaches values close to unity, and the growth occurs then very rapidly.

6. Warm Rain

There are two basic mechanisms by which rain can develop. The first one, sometimes alluded to as the Bowen–Ludlam or 'warm rain' process, involves only liquid phase, with growth first by condensation and then by collection. This is undoubtedly the mechanism operating in tropical clouds warm enough not to contain ice in their entire volume, and producing rain. It probably operates in other more frequent cases as well.

The theory simply assumes that some drops grow larger than others (right hand tail of size distribution spectrum), start falling at an appreciable velocity with respect to the smaller ones, until growing to raindrop size ($> 100 \, \mu m$) by collecting them.

Thus the whole process of growing from the drop produced by an activated nucleus to a raindrop is a combination of growth by condensation first, and then by collection. If we differentiate Equation (5) we see that the rate of growth by condensation is of the type

$$\frac{dR}{dt} = \frac{C}{R} \tag{9}$$

where we now call R the radius of the growing drop. If we combine this with Equation (7), we can represent schematically the growth rate dR/dt as a function of R by a graph of the type of Figure 13. Here the thick curve is meant to combine the growth by condensation, which would be of the type of Equation (9) and be responsible for the left branch, and the growth by collection, given by (7), responsible for the right branch. In the intermediate region the combined growth by both mechanisms has a very low rate, with a minimum around 15 to $20 \, \mu m$. How low is still a debated question*, but this size range for the growing drop gives in any case the bottleneck of the whole process.

Once a drop has grown up to millimeter size, another mechanism limits the growth:

---

* Turbulence and perhaps electric forces may enhance the growth rate.

Fig. V-13. *Combined rate of growth*. The thin lines represent the rate of growth by condensation alone and by accretion alone. The thick line is the combined rate. The figures, given as a general indication, correspond to an idealized simple example in which $f(T, p) = 10^{10}$ s/m^2, $(S - 1) = 0.05\%$ and the growth by accretion occurs by collecting droplets of uniform $10\,\mu$m radius.

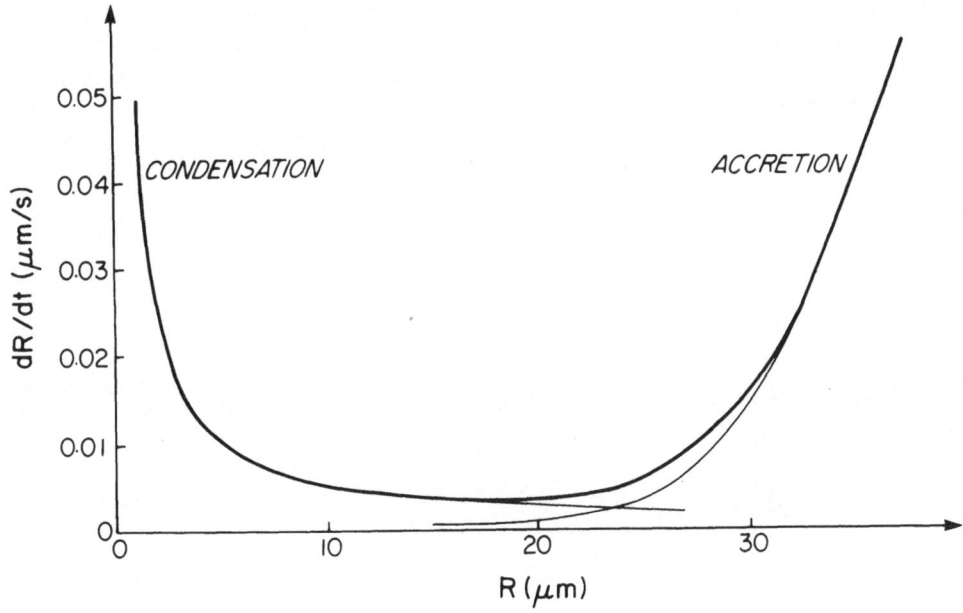

collisions with smaller drops lead to disruption into several fragments. This keeps the size of raindrops below 2 to 3 millimeter radius, until they fall out of the cloud and down to the ground, as rain.

7. Ice Formation

We have seen that spontaneous nucleation of water vapour into liquid droplets does not occur in the atmosphere. Nor does spontaneous nucleation into ice crystals (spontaneous sublimation) either.

Spontaneous nucleation of water into ice does occur, but only when the cloud droplets reach a temperature of $\sim -40°$C. This temperature is slightly dependent on the drop size, and rather numerous laboratory experiments have been performed to study this problem.

Thus, when a cloud parcel rises, it only becomes entirely glaciated at about $-40°$C. Between $0°$C and $-40°$C no spontaneous nucleation occurs. However, freezing can proceed by another mechanism, which we shall now consider.

A certain number of the insoluble aerosol particles (or containing an insoluble core) have the property of facilitating the initiation of ice phase in water. These are called *freezing nuclei* or *ice nuclei*. From the cloud physics point of view, it is very important to know how many ice nuclei per unit volume become active at each temperature. This is done with *ice nuclei counters*. In these, saturated air is cooled (by introduction into a refrigerated chamber or by adiabatic expansion) to a certain temperature, and the number of ice crystals formed is counted by various techniques. Figure 14 summarizes the results obtained by several authors in different locations. It may be seen in the figure that the concentrations of ice nuclei active at temperatures above $-30°C$ are only a small fraction of the particles present in the atmospheric aerosol. It can also be seen that more and more become active as the temperature is decreased.

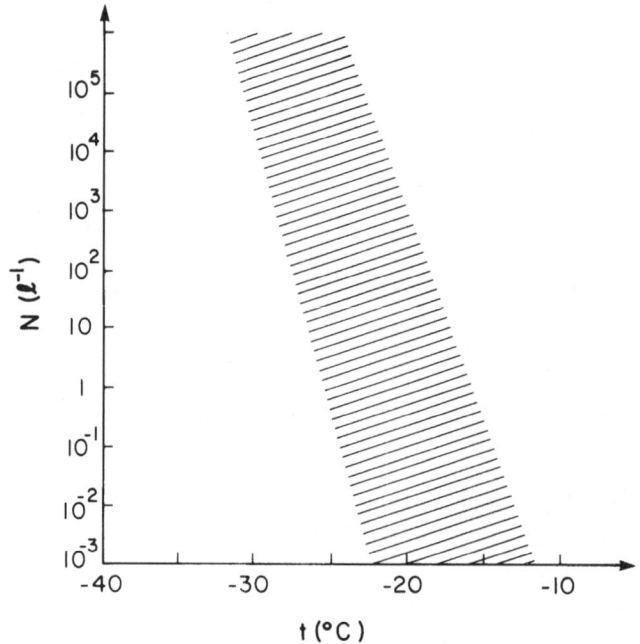

Fig. V-14. *Ice nuclei in the atmosphere.* N = number concentration (per litre); t = temperature at which the nuclei are measured. Concentrations vary considerably with time and place, but most measurements fall somewhere within or close to the shaded area. The slope of this region gives an indication of how the concentration increases as the activation temperature is lowered. At a given time and place, and using a given technique for the measurement, the values of concentrations and their dependence on temperature are much better defined and more consistent.

The reason why certain particles (and not others) are more or less efficient in favouring the formation of ice lies in the molecular arrangements of their crystal lattices. Their surfaces present a pattern on which water molecules can arrange themselves easily according to the crystal lattice of ice. This property is called *epitaxis*. Some natural solids, e.g. micas, have this property, and then act as ice nuclei. Artificial aerosols can also be produced which are particularly efficient; this is the case of silver iodide, widely used in rain stimulation, as we shall see later.

8. Snow, Hail and Rain by the Ice Process

We come now to another mechanism by which precipitation can develop. In fact, it is generally the mechanism supposed to be active in most cases in temperate climates. It is sometimes called the Bergeron–Findeisen process, from the names of its proponents.

According to Figure 14, a significant number of droplets in a cloud will start freezing at temperatures of -15 to $-20°C$; in concentrations of the order, say, of one per liter.

When this happens, the frozen drop becomes a particle of ice surrounded by water droplets in an atmosphere at water vapour saturation with respect to water. Now let us recall the equilibrium water vapour curves of water and ice; schematically, we have the curves of Figure 15. We see that below $0°C$, the saturation water vapour of ice is less than that of water. The consequence is that our ice particle is now in an environment with high supersaturation (with respect to condensation on ice). This supersaturation depends on

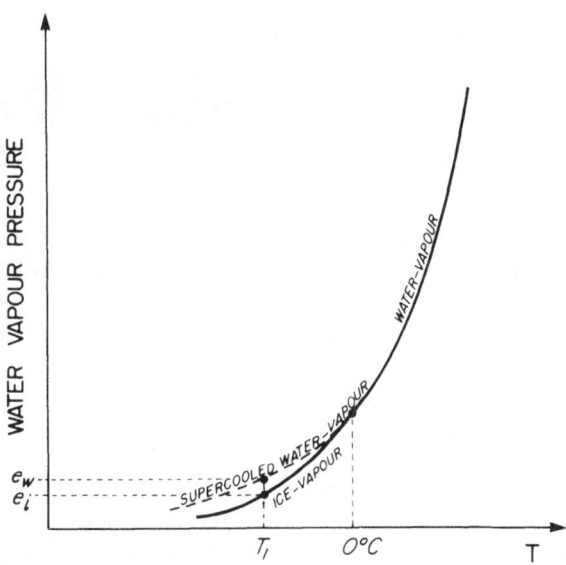

Fig. V-15. *Saturation vapour pressure of water and ice* (schematic; see Fig. II, 1 for plot in scale). At $T_1 < 0°C$, the vapour pressure e_w of (supercooled) water is appreciably larger than that of ice e_i.

the temperature, but may be as high as 30%. The result is that the frozen droplet starts growing by condensation very quickly, and in a few minutes becomes an ice crystal with dimensions of the order of a millimeter. The process implies essentially a distillation of water from the droplets onto the ice particle. We see that the difference with the growth by condensation of a water droplet lies in that now the supersaturation is much higher, which makes condensation a much more efficient process. This rapid growth is illustrated by Figure 16.

113

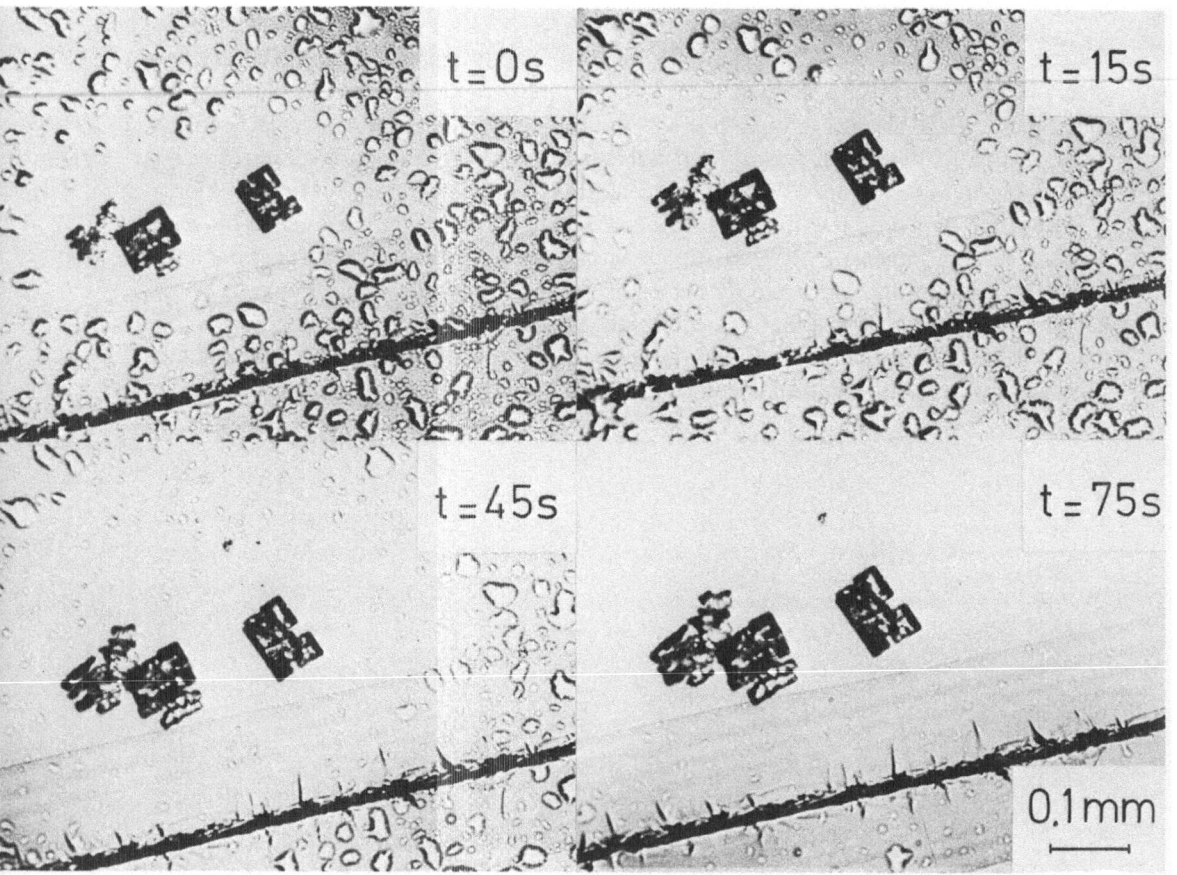

Fig. V-16. *Bergeron–Findeisen effect*. Growth of ice crystals at the expense of surrounding super-cooled water droplets, on a microscope slide. Temperature − 12°C.

Once a crystal of the order of a millimeter is formed, it starts falling with appreciable velocity (of the order of 50 cm/s) in the cloud, collecting the water droplets with which it collides. As the temperatures are well below 0°C, these droplets freeze on contact with the ice. Thus an ice particle is growing quickly. How fast this growth occurs will depend, of course, on the liquid water content (concentration of water in form of droplets, per unit volume) of the cloud. How far it can go depends on how long it remains falling within the cloud, which in turn depends on the updraught currents that maintain the particles in the cloud and on the dimensions of the cloud. Thus, the results may vary widely. We can have snow, consisting of ice crystals often clumped into snow flakes; in this case, growth has been only by sublimation. Or else, we can have small aggregates of frozen droplets, constituting a small soft ice pellet; or, the particle may have the chance of growing much larger into a hailstone. These forms of ice precipitation will be considered in §9. If snow or small ice pellets fall below the 0°C isotherm, they start melting and may arrive to the ground entirely melted, as rain.

In computing the growth of particles, we obviously must take into account the growth formulas, which depend on a number of parameters, like $(S - 1)$, $f(T, p)$, E, w, v. But this is not enough; as indicated above, the trajectory of the growing particle within the cloud is also essential. Not only the previous parameters will vary with the position of the particle in the cloud, but also the possibility of continued growth by remaining within the cloud depends on the trajectory and dynamic factors. Here the updraught velocity U and the thickness Δz of the cloud are essential parameters. Thus a light particle in a strong updraught will be carried aloft and out through the top before it has a chance to grow large. A particle close to the cloud base, in a weak updraught, will soon fall down before growing large. On the other hand, a large particle like a hailstone may grow in a thick cloud starting close to the base and being carried up close to the top, before falling with rapid growth due to its acquired size, while maintained in the cloud during a rather long period of time thanks to a strong updraught.

It is clear from these comments that modelling and computation of precipitation development is a complicated problem where the given conditions — and the results — may vary widely. Without going into any details, we shall point to a few further considerations aimed at illustrating the previous remarks.

Integration of the growth formulas can lead to determination of the size as a function of time: $R = f(t)$. If we now consider the vertical trajectory of the growing particle, we can write

$$\frac{dz}{dt} = U - V \tag{10}$$

If we now write for the growth function $dR/dt = F$ (e.g., Equation (7)) and divide it by (10) we can obtain

$$\frac{dR}{dz} = \frac{F}{U - V} \tag{11}$$

a formula that can be integrated to obtain $R = f(z)$. Equation (11) can be integrated:

$$\int_{R_0}^{R} \frac{U - V}{F} \, dR = \Delta z \tag{12}$$

where V and F depend on R in general and U has been assumed constant for simplicity (not varying with z). If, for instance, we desire to calculate the final size of a drop starting with a radius R_0 close to the base, carried upwards until it becomes big enough to fall within the updraught, and then falling until it comes down from the base as a raindrop, we shall write in (12) $\Delta z = 0$:

$$\int_{R_0}^{R} \frac{U - V}{F} \, dR = 0 \tag{13}$$

and this integral will give the final size R. The trajectory would correspond to a curve of the type of Figure 17, where $z = 0$ corresponds to the base of the cloud. It is obvious that for this process to be possible, the cloud must have a thickness no smaller than Δz corresponding to the top of the trajectory.

The attempt to develop more realistic models than these simple computations implies considerations of the dynamics of the cloud, including such complicated factors as the structure of updraughts and downdraughts and turbulent mixing with the surroundings. This adds very greatly to the complexity of the problem.

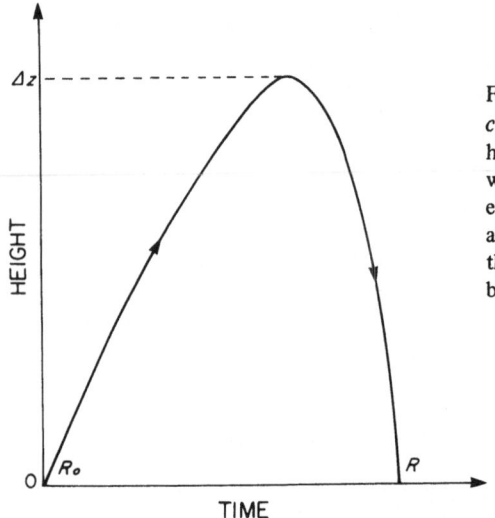

Fig. V-17. *Trajectory of a drop growing within a cloud.* A drop of initial radius R_0 at a reference height 0 (e.g. the base of the cloud) is carried upwards by an updraught while growing. When large enough to overcome the updraught, it will fall back and cross the reference level with a radius R. For this to be possible, the top of the cloud must not be lower than Δz.

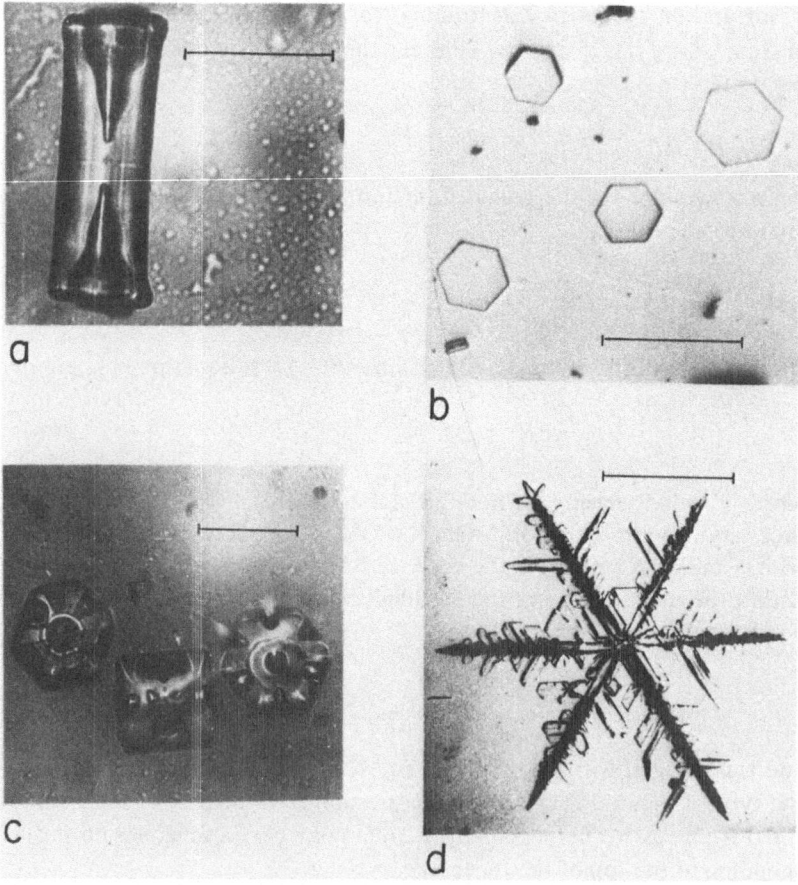

Fig. V-18. *Natural ice crystals.* (*a*) Prism with hollow ends. (*b*) Plates. (*c*) Thick plates or columns, with structure; two lie on their base and one on a lateral surface. (*d*) Star shaped dendrite. The horizontal segment in each photograph measures $100\,\mu$m. All crystals are hexagonal, but their habit depends on the temperature and vapour supersaturation in which they grow.

9. Ice Precipitation

Snow consists of ice crystals grown by sublimation, which can eventually become aggregated into snowflakes. The beauty of snow crystals is well known, and their collection, photographic recording and classification has been the subject of classic books by Bentley and Humphreys and by Nakaya[*]. Ice crystals are always hexagonal, but their forms vary with the temperature and the supersaturation of the cloud atmosphere in which they grow. They can appear as plates, stellar dendrites, prisms or needles (i.e. long thin prisms). Figures 18 and 19 give some examples.

Fig. V-19. *Snow crystals.* The crystals are aggregated into a snow flake. A stellar dendrite is clearly seen, as well as fragments of other crystals.

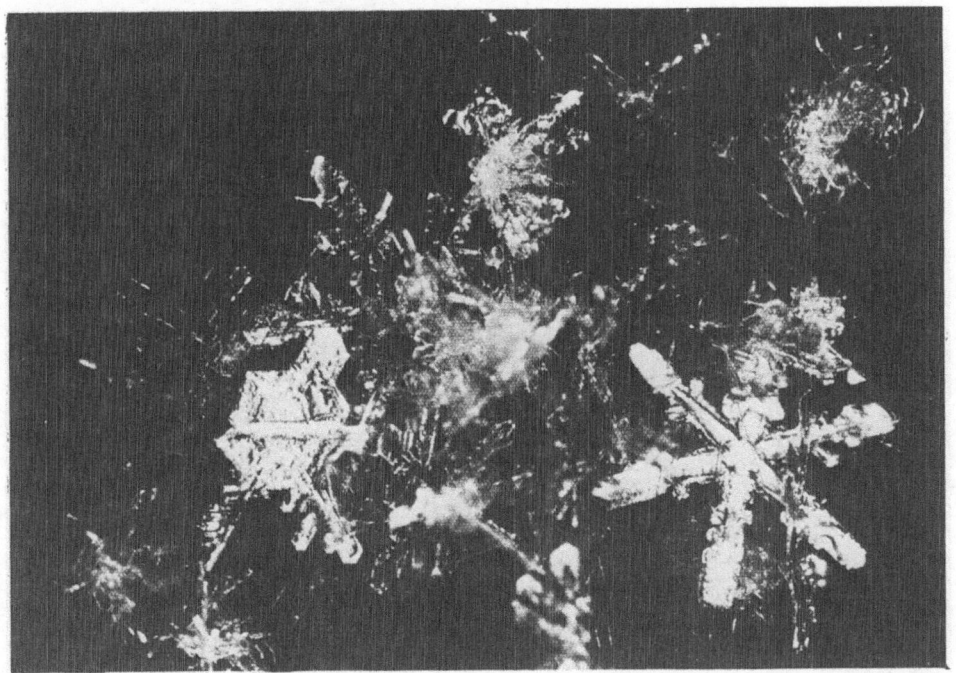

A crystal, falling, may collect supercooled droplets that freeze and adhere to it. This process is called *riming*. After a while, the particle may have grown to a pellet of the size of one to several millimeters mainly formed by frozen droplets with a relatively large volume of air between them, so that it is soft and easy to crush between the fingers. This is called *soft hail* or *graupel*. It often adopts a conical — rather than a spherical — shape. Figures 20 and 21 illustrate this type of precipitation.

A graupel or a frozen raindrop may become the initial stage ('embryo') of a much faster growth, when the conditions are favourable (large updraughts and liquid water

---

[*] W. A. Bentley and W. J. Humphreys, *Snow crystals*, McGraw-Hill, 1931. U. Nakaya, *Snow crystals*, Harvard University Press, 1954.

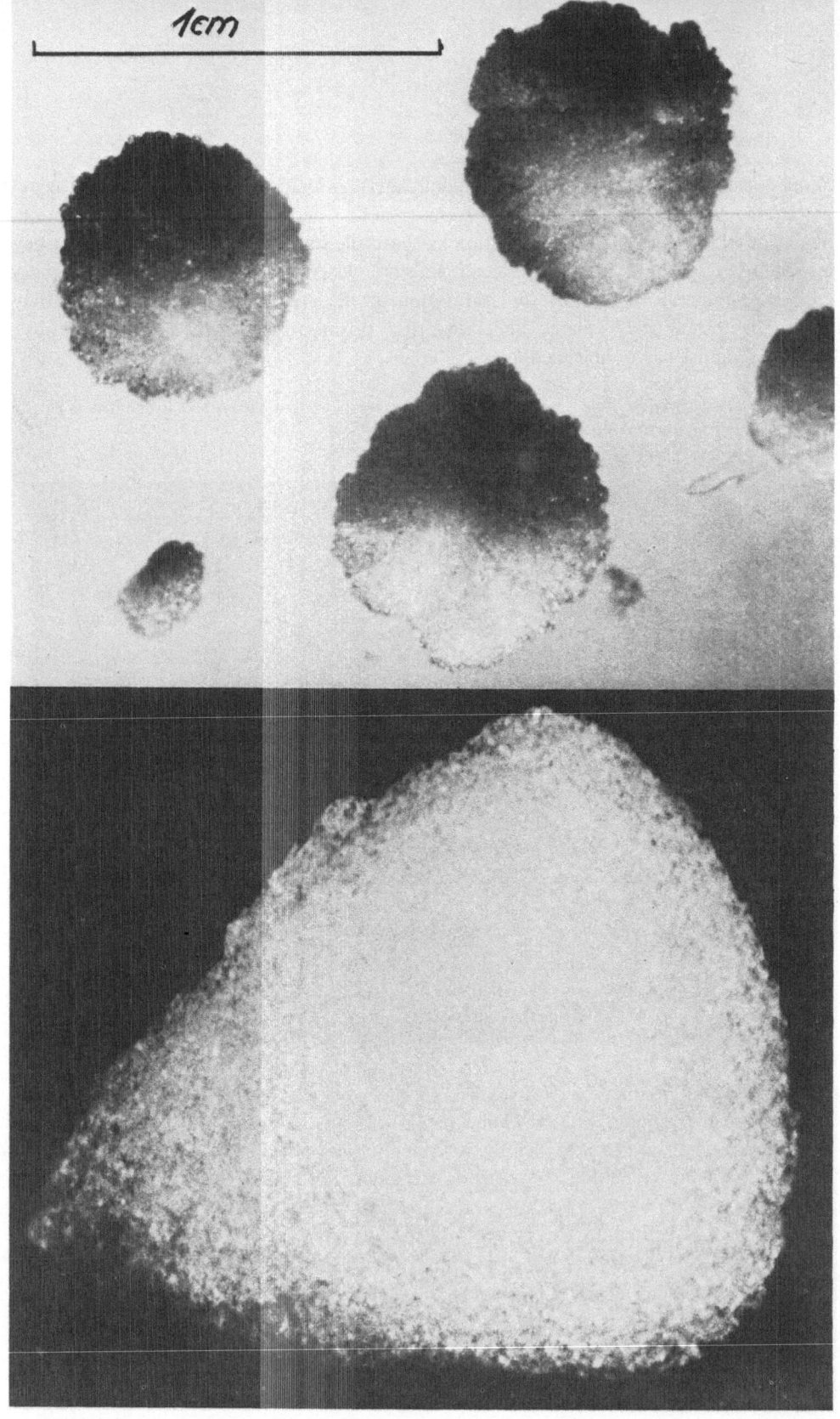

Fig. V-20. *Soft hail or graupel*. Conical graupel with the apex pointing out of the photograph. The rough surface reveals that they consist of aggregates of small frozen droplets.

Fig. V-21. *Graupel*. The picture shows the conical shape and the loose structure of frozen droplets.

Fig. V-22. *Hailstones*. The photograph shows a collection of hailstones gathered in a Petri dish. They show different shapes – rounded, lobed, toroidal or with 'doughnut shape' (no perforation at centre) – and different opacities. The opaque specimens contain large numbers of small air bubbles, giving them a milky appearance. The dish diameter is 15 cm.

Fig. V-23. *Giant hailstone*. The largest hailstone photographed in the U.S.A. Fallen in Kansas on 3 September 1970. Egg and 6-inch ruler added for comparison.

content, as in thunderclouds). This leads to the development of *hailstones*, which may reach surprisingly large sizes (hailstones of diameters larger than 15 cm have been reported). Figure 22 shows some specimens and Figure 23 shows the largest stone photographed in the U.S.A. Hail precipitation occurs in showers, along narrow strips — say of the order of 1 km wide — under the path of the storm. It can produce severe damage to crops, cattle, etc. An interesting feature of hailstones is their layered structure, showing alternate regions of clear and opaque ice, the latter containing numerous air bubbles. This is observed in thin sections with natural light. When the sections are observed between crossed polaroids, the crystal structure becomes apparent. An example is shown in Figure

Fig. V-24. *Cross section of a hailstone*. The figure shows two photographs of thin sections of the same hailstone. The upper one has been obtained with natural light from a section 2 mm thick. It shows the characteristic layered structure of hailstones; the largest bubbles are also visible in the picture. The lower photograph has been taken with polarized light from a 0.3 mm thick section. It shows the crystal structure, with layers of crystals of very different size. The largest diameter of this stone was 4.1 cm.

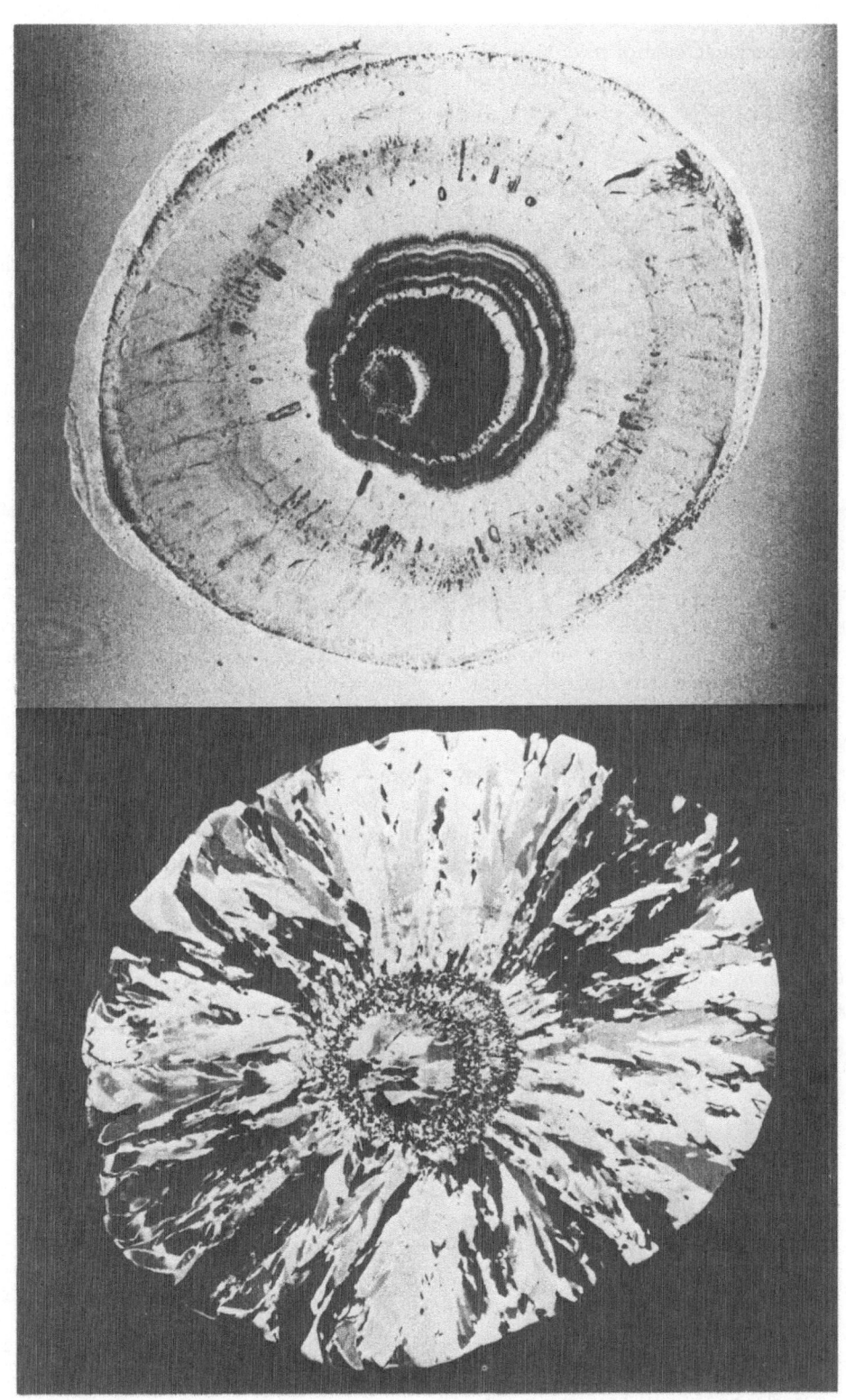

24. These characteristics have been the subject of many studies; their interpretation in terms of the growth process is still a controversial matter.

10. Artificial Modification of Precipitation

The Bergeron–Findeisen mechanism offers the possibility of influencing the process at one critical stage: the appearance of ice particles within a supercooled cloud. This is the basis of most of the experiments made during the last decades (since 1946), in an attempt to modify artificially the amounts or the character of precipitation.

Small scale experiments. From the beginning, it could be seen that success could be achieved in isolated cases. For instance, dry ice (solid CO_2: sublimation point of $-78.5°C$) can be dropped from an aircraft over a supercooled layer cloud, producing glaciation, with subsequent dissipation in a region along the path of the aircraft.

If silver iodide is used (in the form of a thin smoke), its action is based on the fact that its particles start acting as ice nuclei at about $-4°C$, while natural nuclei require lower temperatures. This means that rain stimulation by AgI is restricted to marginal cases where there exist certain basic conditions, such as enough liquid water content, enough development of the cloud to rise to low enough temperatures, etc., which will allow the AgI's action, but would not be sufficient for the natural process. Many experiments have been done comparing the precipitations from pairs of similar clouds, one seeded and one unseeded, and there is no question that the precipitation process can be influenced. The seeding is usually done from an aircraft, using a burner fed with a solution containing AgI.

Large scale experiments. When these techniques are applied to increase rain or snow over an extensive area and for long periods, the problem becomes more complicated. One reason is that evaluation of results is difficult. Because the artificial stimulation is effective only in marginal cases, it is difficult to conclude whether results along a period of time were due to seeding or were natural, and what may be the percentage of increase due to artificial action. Thus, the statistical planning of the experiment becomes essential.

Two resources are employed, in order to make the experiments meaningful. One is the use of control areas. Two areas are chosen, with a long record of precipitation and showing a strong correlation. This is shown in Figure 25, where each point indicates the yearly

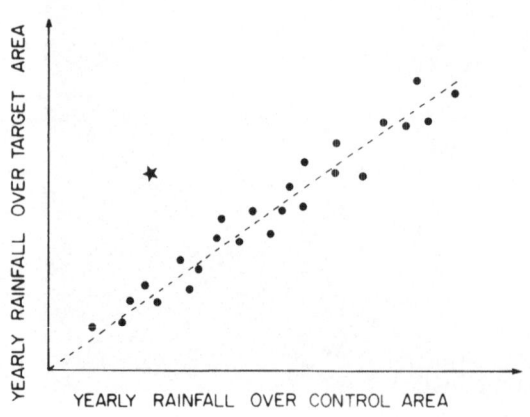

Fig. V-25. *Method of control area.* Each dot corresponds to a given year previous to the experiment. The star corresponds to a (very successful) year of seeding.

122

rainfall in each of the two areas, one of which is to be used for experimental seeding (target area) and the other one to be left as a control. The points in the figure correspond each to one of the previous years, when no seeding was taking place. Now, if we seed during a year, and the rainfalls over the two areas are represented by the star *, we can conclude that the seeding has effectively increased the rainfall over the target area.

Another, more powerful, technique is called *randomization*. It consists in (1) choosing an area and the conditions appropriate for seeding; (2) when these conditions are met, deciding at random whether seeding will be performed or not. All the non-seeded occasions serve as a reference with which to compare the results of the seeded occasions.

Combinations of both techniques can be made. But, in any case, experiments have to be well controlled and run over a number of years (not less than 5), if definite conclusions are to be obtained.

The results are variable, depending on geographic areas, techniques, etc. It is generally accepted that in some cases increases of the order of 10–20% in rain or snow precipitation can be obtained. In other cases, the results are null, or even negative.

Other possibilities of artificial modification of precipitation. Although most of the work was done initially for rain increase, there are other possibilities. The most important, on which costly experiments have been done in the last years in the USSR, USA and other countries, is suppression of hail. Others are: decrease or suppression of lightning in areas where it may lead to forest fires; fog dissipation over airports; production of rain by the warm rain process, using hygroscopic nuclei as seeding material.

Chapter V: Questions

Q1. What is the basic difference between cirrus clouds and the other types of clouds? Between cumulus and cumulonimbus and the other clouds?

Q2. Would you expect to have cirrus clouds at levels of temperature between 0 and $-30°C$? Would you expect to have water clouds within 1 or 2 km of the tropopause? Explain.

Q3. Consider a droplet of pure water in air containing water vapour at the saturation value for the droplet, the system being therefore in equilibrium. Is this a stable or an unstable equilibrium? Explain.

Q4. The decrease of vapour pressure by the presence of a solute (Raoult's law (2)) is called a *coligative* property, the adjective meaning that the property depends on the number of particles – molecules or ions – dissolved per unit volume. Can you mention other coligative properties?

Q5. Consider an atmosphere whose water vapour pressure is exactly equal to the saturation value that you can find in tables. Will that atmosphere be supersaturated, saturated or subsaturated with respect to
(*i*) droplets of pure water, with radius 1 μm,
(*ii*) the plane surface of a solution 1 M of NaCl in water?

Q6. If air may contain many thousands of condensation nuclei per cubic centimeter, why do clouds never contain more than a few hundreds of droplets in the same volume unit?

Q7. What is the difference between a condensation nucleus (CN) and a cloud condensation nucleus (CCN)?

Q8. In which case will rising air reach higher supersaturations: rising at several meters per second, as in a cumulus, or at low velocities (say 10 or 20 cm/s), as in slowly developing layer clouds? Explain.

Q9. How does the terminal velocity of a drop depend on its radius in the range of 10–30 μm? In the range above 3 mm?

Q10. Maritime air is rich in giant nuclei (radius > 1 μm). Would you expect maritime clouds to give rain by the Bowen–Ludlam mechanism more easily than continental clouds? Explain.

Q11. What is a 'cloud condensation nucleus'? An 'ice nucleus'?

Q12. How is it possible that an ice crystal can grow in a cloud to millimeter size in a few minutes by condensation, while a droplet does not grow in that way beyond about $15\,\mu m$?

Q13. Hygroscopic particles have been used to seed clouds, with the purpose of producing rain. On what idea is this based, and what criteria would you apply to choose an appropriate particle size?

Chapter V: Problems

(Any necessary constants not given in the statement of a problem will be found in the Table of Constants on pages x–xi)

P1   Consider a nucleus of NaCl of mass 3×10^{-14} g. Derive:
(a) The radius of a droplet containing this nucleus in solution, for which the vapour pressure e_r' is exactly equal to that of pure water with a plane surface e_s.
(b) The critical radius, over which the nucleus becomes activated.
The two results will be in the order of magnitude of micrometers. You can use this fact to simplify the calculations.
Notice that in calculating molar fractions, each formula weight ('molecular weight' M_{NaCl}) of NaCl counts as two moles, because of its total dissociation in solution. $T = 0°C$.

P2. A water droplet containing 3×10^{-16} grams of sodium chloride has a radius of $0.3\,\mu$m.

(*a*) Calculate its vapour pressure.

(*b*) If the droplet is in equilibrium with the environment, what is the supersaturation (expressed in percentage) of this environment?

(*Note*: In computing molar ratios, each formula weight ('molecular weight' M_{NaCl}) of NaCl must be considered as 2 moles, because of the total dissociation in solution.) The temperature is 25°C. At that temperature, the saturation vapour pressure is $e_s = 31.67$ mb.

P3. A drop of pure water is growing by condensation in an atmosphere kept at 0.2% supersaturation of water vapour. The function

$$f(T,p) = \frac{RT\rho}{DMe_s} + \frac{L_v\rho}{KTM}\left(\frac{L_v}{RT} - 1\right)$$

(where R = gas constant, T = absolute temperature, p = pressure, ρ = density of liquid water, D = diffusion coefficient of water vapour in air, M = molecular weight of water, e_s = saturation vapour pressure at T, L_v = molar heat of vaporization, K = thermal conductivity of air) is assumed to remain constant, with a value 10^{10} s/m². For the present calculation, the approximation is made of neglecting the effects of curvature and ventilation.

(*a*) Plot dR/dt (where R = drop radius and t = time) as a function of R, between $R = 2\,\mu$m and $R = 20\,\mu$m.

(*b*) How long does it take for the drop to grow (*i*) from $R = 2\,\mu$m to $R = 5\,\mu$m; (*ii*) from $R = 5\,\mu$m to $R = 10\,\mu$m; (*iii*) from $R = 10\,\mu$m to $R = 20\,\mu$m?

P4. A sphere moving in viscous flow through a fluid experiences a drag force given by Stokes' law:

$$f_{drag} = 6\pi\eta r v$$

where η = viscosity of the fluid, r = radius of the sphere, v = velocity. Derive an expression $v = v(r)$ for the fall in the air of droplets not exceeding about $30\,\mu$m radius, a process for which the air flow is viscous.

P5. A drop grows by accretion as it falls through a cloud of uniform droplets of radius $r = 10\,\mu$m, at about 900 mb and 0°C. A set of estimated collection efficiencies and the terminal velocities of the growing drop are given in the following table, for different values of the radius R:

$R\ (\mu m)$	E	$V\,(cm/s)$
30	0.17	9.0
40	0.45	18.7
60	0.56	36.1
80	0.62	56.0
100	0.69	74.7

The terminal velocity of the cloud droplets is 1.3 cm/s. The liquid water content of the cloud is $2\,g/m^3$.

(a) Compute the growth rate dR/dt between $R = 30\,\mu m$ and $R = 100\,\mu m$, and plot it as a function of R.

(b) Plot dt/dR in the same graph and make a graphical estimate of the time taken by the drop to grow from 30 to $100\,\mu m$ radius.

P6. Assume that a cloud drop grows from an initial radius of $1\,\mu m$ in the following idealized conditions: $(S - 1) = 0.05\%$; $f(T, p) = 2 \times 10^6\,s/cm^2$ (cf. formula (3) in text); apart from the growing drop, the cloud is formed by uniform drops of $10\,\mu m$ radius, amounting to a liquid water content of $w = 2\,g/m^3$. Cloud droplets of $10\,\mu m$ fall at 1.26 cm/s. The terminal velocity V of the growing drop and a set of estimated collection efficiencies E with respect to the $10\,\mu m$ radius drops at 900 mb are given in the following table, where R is the radius.

$R\,(\mu m)$	$V\,(cm/s)$	E
15	2.9	0.012
20	5.1	0.023
25	7.6	0.054
30	10.9	0.17
40	18.7	0.45
60	36.1	0.56
80	56.0	0.62
100	74.7	0.69
200	168.0	0.78
400	336.0	0.87
1000	650.0	0.88
2500	910.0	0.83

(a) Calculate the growth rate dR/dt by the two processes (condensation and accretion) and the combined value, and plot the values as functions of R.

(b) Using the graph thus prepared, make a graphical estimate of the time taken by the drop to grow from 10 to $25\,\mu m$.

P7. Assume that a cloud has uniformly 2 grams of liquid water per cubic metre of air in the form of droplets $10\,\mu m$ in radius. A drop of $150\,\mu m$ radius, initially close to the base, is carried upwards in an updraught of 10 m/s, while growing by accretion. When its size is large enough, it falls back down and out of the cloud.

(a) What is the final size?

(b) What would be the minimum thickness of the cloud, for this process to be possible? Assume that for the relevant drop sizes and cloud height, the falling velocity can be approximated by the relation $V \simeq 8.8\,R$, where R is the radius of the drop in mm and V is given in m/s. The approximations $V \gg v$ and $[1 + (r/R)]^2 \simeq 1$ can also be made (where v is the falling velocity and $r = 10\,\mu m$ is the radius of the cloud droplets). The collection efficiency E can be approximated by an average of 0.8.

VI. Atmospheric Electricity

1. Electric Properties of the Atmosphere

There normally exists a vertical electric field in the atmosphere. A simple experiment can prove this. Assume that you have a conducting body, as in Figure 1a, held through an insulator. If there is a field **E**, there will be a polarization of charges, as indicated in the figure. Now let us connect the body to ground (part b), whereby positive charges flow to ground. Then let us disconnect the body and bring it into a Faraday cage attached to an electrometer (c). The instrument will show that the body is charged, as should be expected from the existence of **E**.

There are also free charges in the atmosphere. This becomes obvious in that a well insulated charged conductor exposed to air gradually loses its charge.

Given the existence of both a field and free movable charges, it is implicit that air possesses some conductivity, and that currents must be present. Thus the description of the atmospheric electric properties cannot be purely electrostatic; inasfar as these properties remain constant, they must describe a stationary state rather than a static one.

Before starting to describe these properties we must make a basic distinction regarding the state of the atmosphere where the properties are studied. We speak of:

Fair weather – when there is no precipitation, less than 4/10 of the sky is covered by clouds, and no other extreme conditions of cloudiness, visibility or wind are present.

Disturbed weather – in other situations. Thunderstorms are typical examples of disturbed weather.

The properties are so different in these two types of situations, that they must be studied separately. We shall start with fair weather conditions.

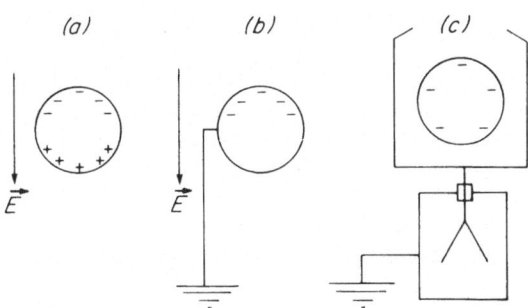

Fig. VI-1. *Existence of an atmospheric electric field.* (*a*) A conductor (sphere) is polarized in the atmospheric electric field **E**, which points downwards. (*b*) Connection to ground will discharge the positive charges to ground. (*c*) If the sphere is now insulated and introduced into a Faraday cage connected to an electrometer, this will show the presence of a net (negative) charge.

2. Fair Weather Electric Field and Space Charge

Measurements of the electric field at the ground E_0 are very variable with place and time (as are all electric properties), but in average the field is vertical, with a magnitude

$$E_0 \cong -120\,\text{V/m} \tag{1}$$

where the negative sign indicates that it is directed downwards.

According to the laws of electrostatics and taking into account that the Earth is a conductor, the presence of this field implies the existence of a surface density of charge, σ, at the ground, given by

$$E_0 = \frac{\sigma}{\epsilon_0} \tag{2}$$

where ϵ_0 = permittivity of free space ($\epsilon_0 = 1/4\pi$ in the electrostatic system of units, and $\epsilon_0 = 8.854 \times 10^{-12}\,\text{F/m}$ in the SI system).

Accordingly, the value of σ is

$$\sigma \cong -3.2 \times 10^{-4}\,\text{e.s.u./cm}^2 = -1.1\,\text{nC/m}^2 \tag{3}$$

(e.s.u. = electrostatic unit of charge; 1 nC = 1 nanocoulomb = 10^{-9} C).

If we now measure E at increasing altitudes, we find that it decreases (in absolute value) very rapidly. This means that there must be a positive net space charge in the air.

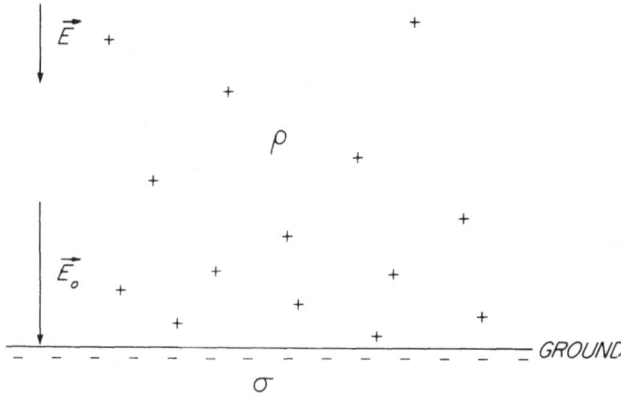

Fig. VI-2. *Space charge in the atmosphere and surface charge at the ground.* The ground has a surface charge of density σ. Above it, the air contains a space charge of density ρ, decreasing with height. The electric field **E** points downwards, is maximum at the ground (**E$_0$**) and decreases with height.

The relation between E and the space charge density ρ, assuming that E is only a function of the height z, which is essentially correct, is given by Poisson's equation:

$$\frac{dE}{dz} = \frac{\rho}{\epsilon_0} \tag{4}$$

The charge density ρ decreases rapidly with z, and correspondingly E decreases more and more slowly. As it does so, it tends to 0; this means that at a high enough level, the total positive space charge between ground and that level compensates completely the

130

negative surface charge at the ground. These relations can be expressed by integrating (4). At each height z, $E = E(z)$ is given by

$$E = E_0 + \frac{1}{\epsilon_0} \int_0^z \rho \, dz \tag{5}$$

At a high enough level (indicated by $z = \infty$):

$$0 = E_0 + \frac{1}{\epsilon_0} \int_0^\infty \rho \, dz \tag{6}$$

Or, introducing (2):

$$\sigma = -\int_0^\infty \rho \, dz \tag{7}$$

Figure 3 shows the type of variation of E with z, and Figure 4, that of ρ given in elementary charges per cm^3 (1 elementary charge $= 1.6 \times 10^{-19}$ C).

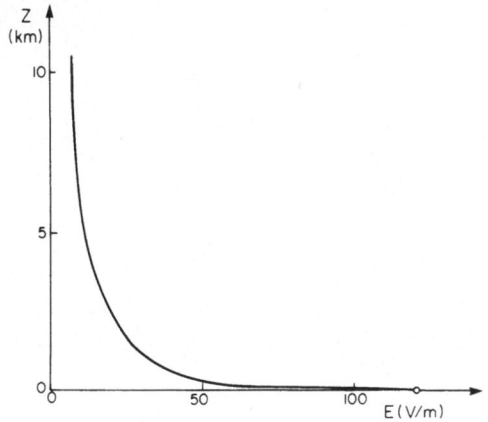

Fig. VI-3. *Variation of electric field with height.* $E =$ electric field; $z =$ height. Conditions may vary widely; the figure shows a typical curve.

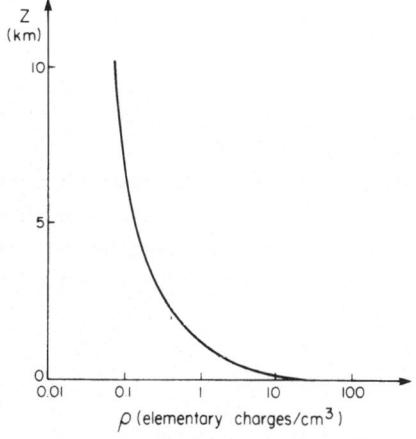

Fig. VI-4. *Variation of space charge with height.* $\rho =$ space charge; $z =$ height. Conditions may vary widely; the figure shows a typical curve. Notice that a logarithmic scale is used for ρ.

131

It remains to consider the variation of the electrical potential V. The Earth is taken as a reference for this parameter, so that $V_0 = 0$. To obtain $V(z)$ we recall that the field is equal to minus the gradient of the potential; in this unidimensional case:

$$E = -\frac{dV}{dz} \tag{8}$$

By integration,

$$V(z) = -\int_0^z E\,dz \tag{9}$$

As E is relatively large and negative close to the ground, and decreases rapidly tending to 0, V must increase rapidly at first, tending to a final constant value at high enough levels. This final value is about $3 \times 10^5\,V$ and is practically reached at 20 km, as indicated approximately in Figure 5:

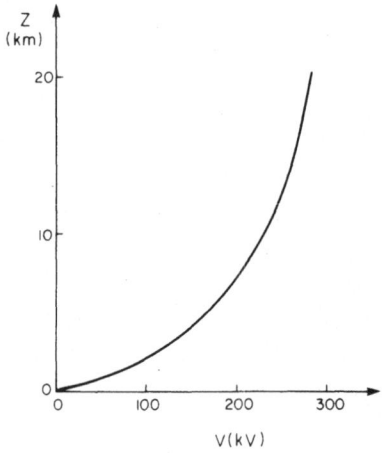

Fig. VI-5. *Variation of potential with height. V* = potential; *z* = height.

3. Atmospheric Ions

Now let us consider the nature of the space charge in the air.

We have seen that in the high atmosphere, the ultraviolet radiation coming from the Sun produces a large amount of ionization of the molecules and atoms, leading to the existence of the ionosphere. In the lower atmosphere such ionizing radiation does not arrive, having been exhausted at higher levels. However, a small amount of ionization is produced by:

— radioactive gases (particularly Rn)
— radioactive substances at the ground
— cosmic rays.

The three contributions are comparable, although the first one is the largest. The production varies with the altitude and latitude, but is of the order of 10 ion pairs/s . cm³.

132

The primary ions produced are molecules of air constituents ionized by the radiation (positive ions) and negative ions produced by attachments of the electrons to other molecules. These primary ions undergo several processes. First of all, they react with neutral molecules and attach water molecules from the water vapour always present, forming *cluster ions*. Most of these are hydrated hydroniums, as indicated in the following scheme of reactions, which is an example for the case of N_2^+ as primary ion:

$$N_2^+ + H_2O \rightarrow N_2H^+ + OH$$
$$\quad \rightarrow + H_2O \rightarrow H_3O^+ + N_2$$
$$\qquad \rightarrow + H_2O \rightarrow H_3O^+ \cdot H_2O$$
$$\qquad \quad \rightarrow H_3O^+(H_2O)_n$$

Net result: $N_2^+ + (n + 2)H_2O = H_3O^+(H_2O)_n + OH + N_2$

This evolution occurs in times of the order of some microseconds. These cluster ions $H_3O^+(H_2O)_n$ are relatively stable, and are supposed to constitute most of the ions of molecular size in the air. However, they can undergo further, slower reactions, attaching other molecules, particularly of trace gases such as NO_2, SO_2, etc.

All these ions, with the size of molecular clusters, are called *small ions*.

Small ions can attach to aerosol particles, giving charged particles. These are called *large ions*.

Small ions can also recombine, positive with negative, thus losing the charges. In an approximate way, a steady state can be formulated, in which as many small ions are produced (by radioactivity and cosmic rays) as disappear by recombination or by attachment to particles:

$$q = \alpha n^2 + \beta n N \tag{10}$$

Here q = total rate of production, per unit volume and unit time, n = number concentration of small ions, N = number concentration of particles, α = recombination coefficient = 1.6×10^{-6} cm^3/s at STP, β = attachment coefficient = 0.4 to 15×10^{-6} at STP.

The total number concentration at the ground is of the order of 100 to 1000 cm^{-3}. There are more positive ions than negative ones, and the difference produces the net positive space charge which we have considered before. For instance, there may be $n_+ = 600$ cm^{-3} positive ions and $n_- = 500$ cm^{-3} negative ions, which gives a net positive charge of 100 elementary charges/cm^3.

The concentration of large ions is smaller than that of small ions in pure air from non-polluted areas (such as over the oceans), but may be much more numerous in polluted areas, as a result of the capture of the small ions by the numerous aerosol particles.

4. Conductivity

The presence of mobile charged particles imparts a certain conductivity to the air. The conductivity increases with height, owing mainly to the decrease in air density. Also, the number concentration of small ions increases in average with height; notice that this is not inconsistent with the decrease in net space charge; it simply means that the total number concentrations of both positive and negative ions increase, but their difference decreases with height. Figure 6 gives an indication of the variation of conductivity λ with height*. The conductivity λ is defined as the proportionality constant in the formula

$$\mathbf{j} = \lambda \mathbf{E} \tag{11}$$

where $\mathbf{j}$ is the current density produced by the field $\mathbf{E}$. λ is measured in s^{-1} in the electrostatic system of units and in $\Omega^{-1}\,\text{m}^{-1}$ in the SI system ($1\,\Omega^{-1}\,\text{m}^{-1} = 9 \times 10^9\,\text{s}^{-1}$).

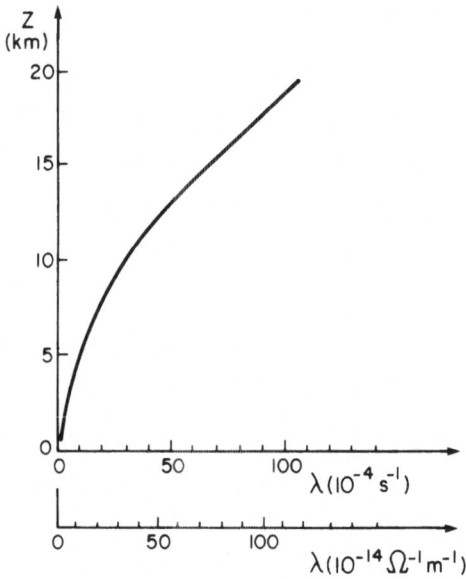

Fig. VI-6. *Variation of air conductivity with height.* λ = conductivity; z = height.

In formulas (6) and (7) we wrote $z = \infty$ as the upper limit of integration. In fact, ρ and E become ~ 0, and V a constant, already at levels of the order of 20 km. If we go to higher altitudes, the air becomes more and more conductive. This is particularly so when we arrive at the ionosphere. For all practical purposes, above about 65 km (approximately the base of the D region) the atmosphere can be considered as a conductor.

From the existence of $\mathbf{E}$ and of a finite λ, we conclude that there must be present a vertical, downward, current density $\mathbf{j}$ which must be roughly constant through the troposphere, for continuity reasons; i.e. as the height increases, $\mathbf{E}$ decreases, but λ increases, keeping $\mathbf{j}$ constant.

5. The Fundamental Problem of Atmospheric Electricity

We can now summarize the electrical situation in fair weather in the following way. We have two concentric spherical conductors: the Earth and the ionosphere (see Figure 7). Between the two, there is a semi-insulating layer of atmosphere, with a small conductivity λ. The

* This is a very small conductivity. For comparison, the conductivities of copper, germanium (semiconductor) and glass at room temperature are in the orders of 10^8, 10^{-1} and $10^{-11}\,\Omega^{-1}\,\text{m}^{-1}$, respectively.

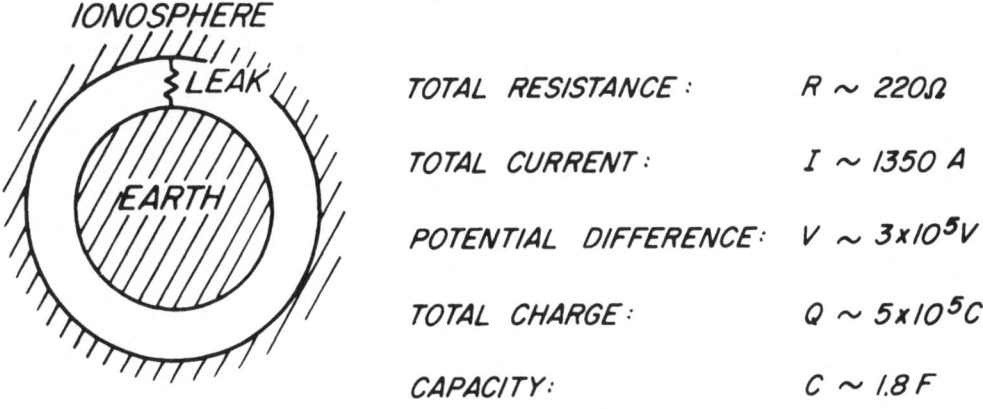

TOTAL RESISTANCE : $R \sim 220\,\Omega$

TOTAL CURRENT : $I \sim 1350\,A$

POTENTIAL DIFFERENCE: $V \sim 3 \times 10^5 V$

TOTAL CHARGE : $Q \sim 5 \times 10^5 C$

CAPACITY: $C \sim 1.8\,F$

Fig. VI-7. *Earth-ionosphere as a spherical capacitor with a leak*. Both ground and ionosphere are conductors. Between them mobile charges (ions) provide a leak.

Earth surface can be considered as one of the plates of a spherical capacitor. The other 'plate' is a diffuse region containing space charge of maximum density close to the Earth surface and decreasing with the distance from it. These charges are imbedded in a dielectric (air) which becomes less and less insulating and more conducting with the distance, until it finally becomes a conductor (ionosphere). This spherical capacitor is charged at a total potential difference of 300 kV. The space charge in the intermediate region (the charge of the second plate) is not static, but is moving toward the Earth, constituting a leaking current.

It can be calculated from (11) that the current density is $j = 2.7 \times 10^{-12}$ A/m². As the total surface of the Earth is $S = 5 \times 10^{14}$ m², the total current discharging the capacitor is of the order of $I \cong 1350$ A.

The total charge Q in the capacitor is

$$Q = S\sigma$$
$$= 5 \times 10^{14} \times 1.1 \times 10^{-9} = 5.5 \times 10^5 \, C \qquad (12)$$

and its capacitance C is*

$$C = Q/V = 5.5 \times 10^5/3 \times 10^5 = 1.8 \, F \qquad (13)$$

The total equivalent resistance R of the atmosphere can be obtained from:

$$R = V/I = 3 \times 10^5/1350 = 222 \, \Omega \qquad (14)$$

Therefore, the charge Q should disappear with a time constant τ:

$$\tau = RC = 407 \, s \cong 7 \, min$$

Thus it is evident from the previous consideration that in order to have a steady state, some powerful mechanism must be continuously charging the atmosphere with respect to the Earth, i.e. transporting positive charges, from Earth to atmosphere, at the rate of ~ 1350 A. This is sometimes referred to as the fundamental problem of atmospheric electricity.

* According to this capacitance, an equivalent spherical capacitor with two conventional conducting plates (such as metal plates) would have a separation of 2.5 km between the plates.

The fundamental hypothesis is made, in order to explain this situation, that the charging process is the thunderstorm activity all over the world. Each thunderstorm acts as a generator, as indicated schematically in Figure 8:

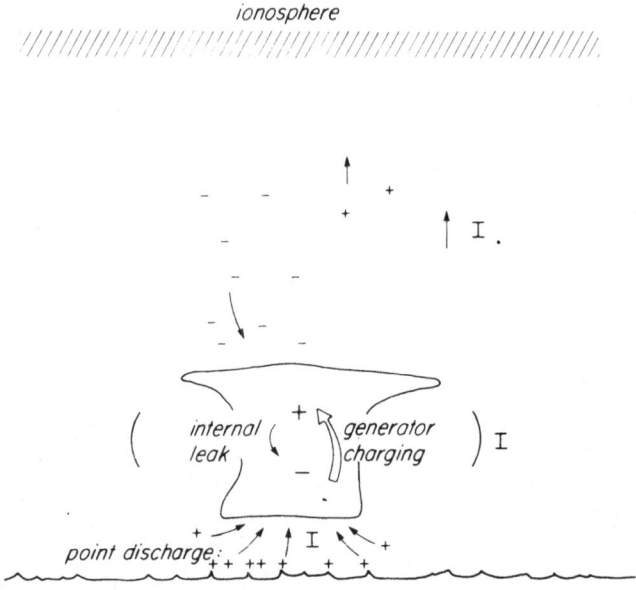

Fig. VI-8. *Thunderstorms acting as generators charging the Earth-ionosphere capacitor.* See text for explanation.

Here we have:

— A charge separation mechanism provided by the thunderstorm, which usually has, as we shall see, positive regions aloft and negative regions at lower levels. This is equivalent to a charging current pointing upwards. An internal leaking current, depending on the conductivity within the cloud (which is not well known), should be subtracted from the charging.

— Above the cloud the field is reversed with respect to fair weather values, due to the charges in the thunderstorm. As the free charges (of both signs) increase with height, the result will be a current directed upwards, mainly carried by downwards migrating negative charges.

— Below the cloud the field is also reversed. Here the current will be both by migration and by convection (through the strong updraughts). To the normal source of charges a very important source of *positive* charges is added by the increased point (corona) discharge at the ground.

Thus the total effect is an upward charging current I, from ground to ionosphere. This will be valid for each thunderstorm, so that the sum of all thunderstorms occurring simultaneously over the Earth constitutes the charging mechanism that maintains the spherical capacitor Earth–ionosphere charged.

If this theory is correct, there should be a parallelism between the world thunderstorm activity and the field E in fair weather regions. This is found to be the case if average values over long periods are calculated for the daily variation of this parameter ('universal daily variation') and of thunderstorm activity. However, this is not so for single days or averages for short periods; the situation is not yet well understood.

6. Thunderstorm Electricity

Cloud physicists have been concerned for a long time with the explanation of thunderstorm electricity, but it still is a controversial and not well understood subject. The cause of this situation is partly the lack of enough basic information, as electrical measurements in storms are scarce, and difficult and costly to obtain.

We shall only mention the most widely accepted explanations. Some researchers believe that the separation of electrical charges within a thundercloud is the result of *convective* currents. Most researchers, however, are in favour of the so-called *gravitational separation* theories, to which we shall refer.

The most frequent distribution of charges within a thundercloud is as shown in Figure 9. The main charges are sometimes inverted. Their magnitude is in the order of tens of Coulombs. When the fields reach large enough values (in the order of thousands of V/cm), lightning strokes occur, both within the cloud and between cloud and ground. In average, some 20 C are neutralized in a stroke.

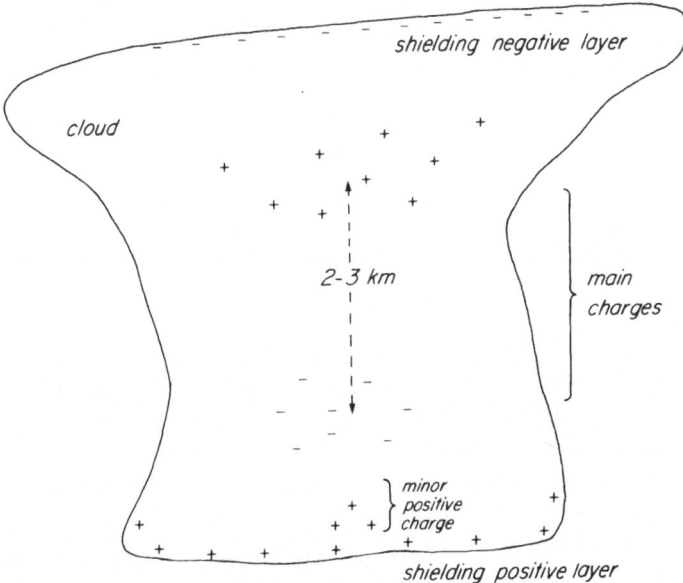

Fig. VI-9. *Electrical structure of a thundercloud*. The main charges are most commonly as indicated in the figure, but sometimes the signs are inverted. The main charges amount typically to several tens of Coulombs, with the negative region at around -5 to $-10°C$ and the positive charge 2–3 km above. A minor positive charge (a few C) is sometimes observed at $0°C$ or slightly higher temperature. Some shielding develops in the surface layers of the cloud, where droplets or crystals capture incoming ions moving along the field towards one of the main charge centres.

Gravitational theories assume that some microphysical process separates charges, leaving the negative ones in heavier particles (raindrops, graupel, hail pellets), while the positive ones go to lighter particles (cloud droplets, ice crystals, ions). The former fall down the cloud and somehow release the negative charge in the lower regions, while the latter are carried by the updraughts to form the upper positive centre.

As to the microphysical processes responsible for the charge separation, at present the following ones seem to be the most probable.

Collisions between hail pellets (or graupel) and ice crystals. The hail pellet is polarized under the influence of the external field E, so that an excess positive charge will lie in its lower part, and negative charges will be in the upper part (see Figure 10).

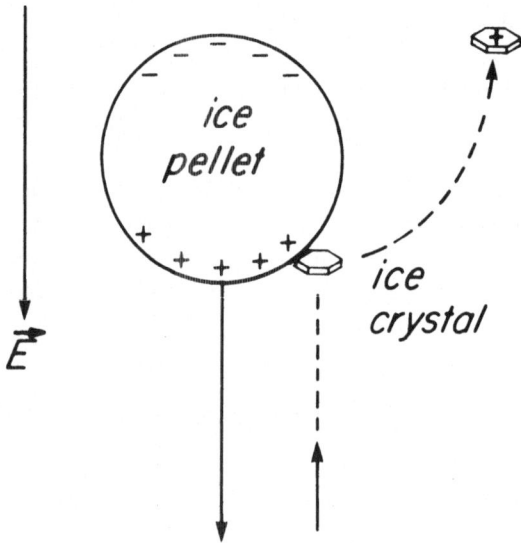

Fig. VI-10. *Charge separation mechanism.* A falling ice pellet is polarized by the external field E. Ice crystals colliding with it acquire positive charge and are carried aloft by the updraught. The complementary negative charge is carried downwards in the pellet.

As the pellet falls within the cloud, it will collide with ice crystals, which we assume present. The collisions will happen at the lower surface, and some of the positive charges may be transferred (under the influence of the external field) from the pellet to the crystal, leaving the compensating excess of negative charge in the pellet. As the light crystals are carried upwards by the updraughts always present in such clouds, while the heavier pellets fall down, a negative space charge builds up in the lower regions of the cloud, and a positive space charge builds up in the upper regions.

Riming of hail pellet or graupel. We recall that by riming is understood the growth of ice at the expense of supercooled droplets that collide with it and freeze onto the surface.

If the droplets colliding with the pellet would remain in their entirety as frozen additions, there would be no charge separation. However, this separation can occur due to two mechanisms:

138

(1) Some droplets, if colliding with enough velocity on the sides of the lower surface (grazing collisions), can splash, producing fragments that fly away. There is in that case a charge separation, ascribable to an electrokinetic phenomenon involving the electrical double layer of ions present at the ice–water interface.

(2) Due to the presence of the external field, the same type of transfer induced by polarization mentioned before will occur here as well. Therefore any drop fragment splashed away, or even whole droplets that make only a grazing collision and bounce rather than stick, will carry positive charge. The complementary negative charge remains, as before, in the pellet (see Figure 11).

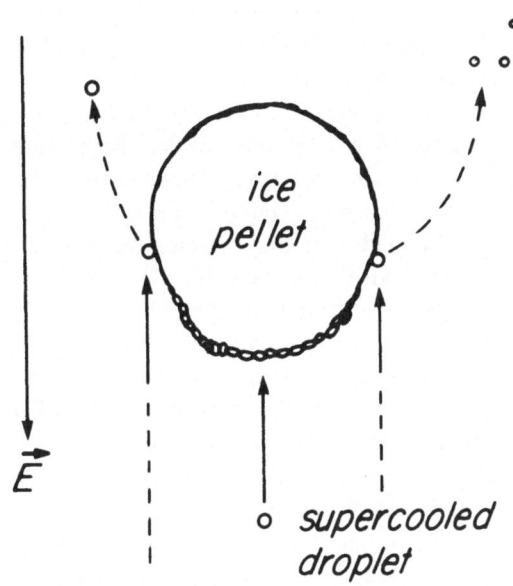

Fig. VI-11. *Charge separation mechanisms.* A falling ice pellet collects supercooled droplets freezing onto it ('riming'). Some droplets in grazing collision break and charged fragments separate; others just bounce (see text for explanation of charging). The positively charged droplets or fragments are carried aloft by the updraught. The complementary negative charge is carried downwards in the pellet.

In the two types of processes described, as the charge separation proceeds, the original field becomes reinforced, i.e. increases its downward value. This, in turn, increases the inductive effects. Thus the increase of the field becomes self-accelerated.

These processes have been studied in the laboratory; estimative calculations have been made for their contributions to thunderstorm electricity. They could, in principle, explain it.

Many other electrification phenomena occur in a mature thundercloud. Some will produce an opposite effect. Others will also contribute to the charge separation. However, the main problem is to explain the initial rise of the electric field within the cloud and its quick growth until lightning can occur.

7. Lightning

We have considered possible mechanisms that lead to the formation of large space charges in a cumulonimbus. The progress of this process is limited not only by the discharging currents from ground to cloud, between the charge centres inside the cloud and from cloud towards ionosphere, but also by electric breakdown with subsequent flashing discharge. These transient, high electric discharges are called *lightning*.

Lightning can occur between cloud and ground, between cloud and the surrounding air or inside the cloud. It may also occur on other occasions when high enough space charges are developed: snowstorms, sandstorms, eruptions from volcanoes. We shall restrict our attention to their most frequent appearance in convective clouds (cf. Ch.V, §1), and in particular to the cloud-to-ground discharges. Examples of lightning are shown in Figures 12a, b. Due to the sudden heating and expansion of the air along the electric discharges, sound waves are produced, which are perceived as *thunder*. Thus electric clouds or storms are also referred to as *thunderclouds* or *thunderstorms*.

Although the lightning appears to the naked eye as a single discharge, it really occurs in steps, in a rather complicated manner. A photographic device, Boys' camera, gave the means to elucidate this structure. This is a camera with an arrangement of rotating lenses that produce a rapidly shifting image on the photographic film; thus a plot of the luminous image or images along time is obtained, revealing the discrete strokes composing a single lightning flash. Stationary lenses with a rotating film have also been used. An example of a photograph taken with this type of camera is given in Figure 13.

The cloud-to-ground flash starts by a weakly luminous discharge that propagates from cloud (usually from the lower negative space charge centre), to ground in steps of the order of 50 m, with pauses between steps of about 50 μs (*stepped leader*). The luminosity is only observable during the advancing periods ($\sim 1\,\mu$s). At the start of a new step the leader sometimes branches and then the branches themselves continue downward in a series of steps which may also branch (see Figure 14). Thus the zigzag features of the lightning channel are due to the hesitant progress of the stepped leader searching for the most favourable path. After some 20 ms[*], the propagation (average velocity of about 10^5 m/s) brings the tip of the leader close to the ground. This discharge is called the stepped leader, and is plotted schematically in the first part of Figure 15, which summarizes the whole process of a lightning flash.

When the tip of the stepped leader has reached a short distance from the ground (say 10–20 m), the electric field at the ground underneath it has become high enough (see below) for a new discharge (*connecting discharge*) to be initiated from some pointed object (tall building, tree, lightning rod). This goes to meet the tip of the stepped leader and continues further upward through the ionized (conductive) channel already produced. This is now a much stronger discharge, carrying peak currents of the order of 10 000 A, propagating at speeds of the order of 10^7–10^8 m/s and strongly luminous. It reaches the cloud in about 100 μs. It is the *return stroke* or *main stroke*.

The lightning may have finished with the return stroke ('single-stroke' flash). More frequently, a 'multiple-stroke' flash will occur. In that case, some hundredths of a second later a new discharge is produced from cloud to ground, draining charges from a region

---

[*] All of these figures are intended as a general indication, as they may vary rather widely from case to case.

(a)

Fig. VI-12. *Lightning.* (*a*) Cloud-to-ground strokes. (*b*) Ground-to-cloud lightning, starting from a tower at Mount San Salvatore, Switzerland.

(b)

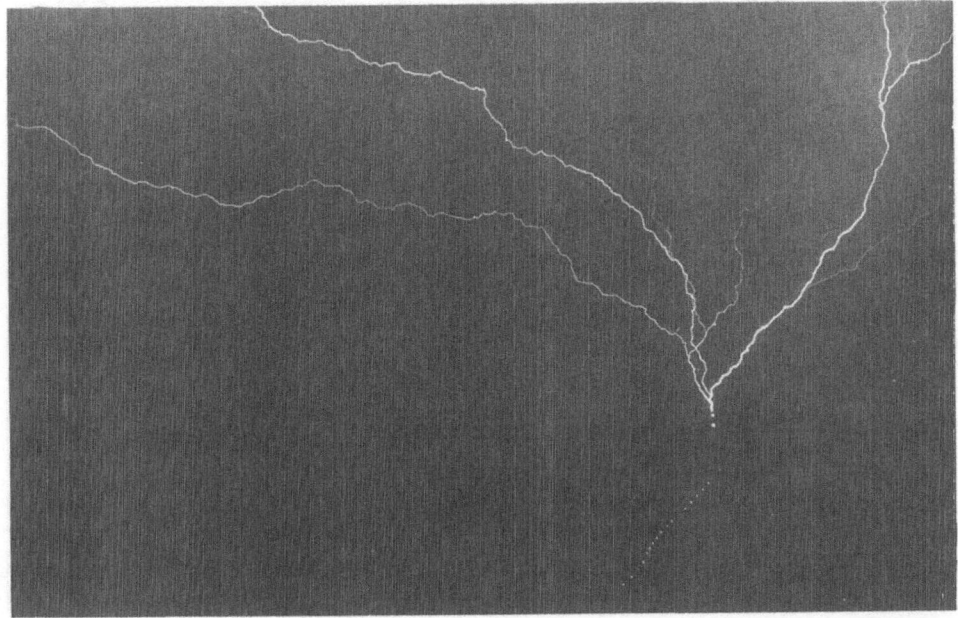

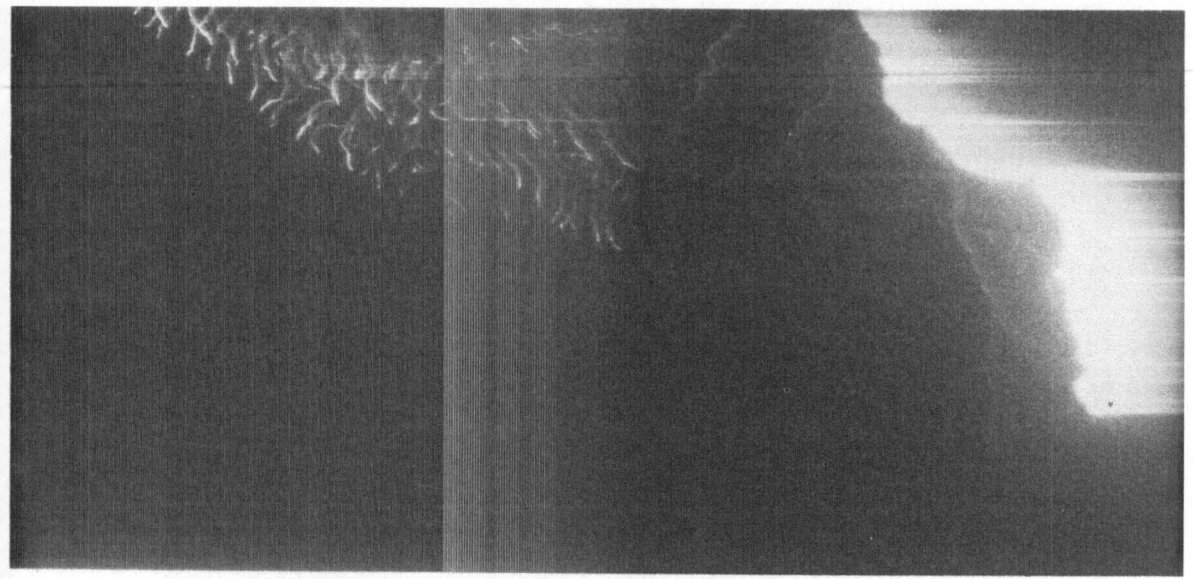

(a)

Fig. VI-14. *Stepped leader and connecting discharge.* (*a*), (*b*) and (*c*) show successive stages of the downward advancing and branching stepped leader in a cloud-to-ground discharge. (*c*) shows the connecting discharge rising from a tree to meet the tip of the stepped leader and initiate the return stroke.

(b)

Fig. VI-13. *Lightning.* (*a*) (left) Photograph of a cloud-to-ground lightning on moving film (film velocity: 30 m/s). The advance downward of the step leader may be seen at the left (time increases toward the right). The luminous discharge at the right is the return stroke. (*b*) (right) Photograph of the same lightning, taken with the usual type of camera. The lightning fell on a tower at Mount San Salvatore, Switzerland.

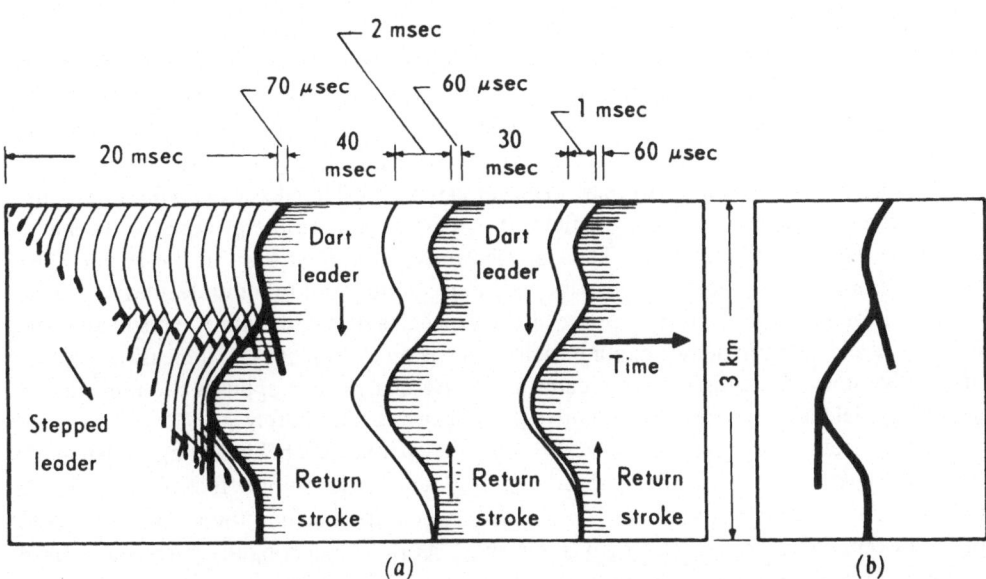

Fig. VI-15. *Lightning structure.* Schematic representation of the different stages of a cloud-to-ground multistroke discharge, as would be recorded by a moving film camera (left; time scale increasing toward the right) and the same lightning as recorded by a conventional camera or as seen by the naked eye (right). The time intervals indicated are orientation figures. The time scale in the left figure has been distorted for convenience. (Copyright © 1969 by McGraw-Hill Book Company.)

close to the origin of the stepped leader. But now it will propagate swiftly and continuously through the still ionized channel at about 2×10^6 m/s (i.e. about ten times faster than the stepped leader, but perhaps twenty times slower than the return stroke), so that it arrives close to the ground in about 2 ms. This *dart leader* is followed by a second return stroke, in the same fashion as after the stepped leader. This whole sequence-interval of a few hundredths of a second, dart leader, return stroke, repeats itself for a variable number of times, usually a few times but occasionally as many as 20 times or more, after which the ionized channel decays and the lightning flash is finished. Sometimes a return stroke is followed by a *continuing current* (~ 100 A) for a period of a few tenths of a second. This occurs after one or several strokes in about half of the lightning flashes.

The total duration of the lightning flash is usually of several tenths of a second. The process of lightning has been described for discharges starting in the cloud, with the initial stepped leader travelling downwards. When very high structures provide a prominent point at the ground (e.g., lightning rod at the top of the Empire State Building in New York), the discharge is often initiated at that point, and the stepped leader travels (and branches) upwards (see Figure 12b).

The structure of lightning, as explained, has been determined for cloud-to-ground or ground-to-cloud discharges, which can be clearly photographed. Intracloud discharges, however, are more numerous; in fact, they are the most frequent type of lightning. A charge of the same order of magnitude as that transferred in a cloud-to-ground discharge is neutralized in them, with a similar time duration.

The theory of the lightning processes is understood in a general way, but many details remain yet uncertain or controversial. The triggering of the stepped leader is probably connected with breakdown starting at pointed ends of water drops elongated by the electric field reaching particularly high values in certain regions of the cloud. Once initiated, the discharge propagates by a process called the *electron avalanche*. The discharge has started by a certain ionization producing positive ions and electrons. Both types of particles will move by action of the electric field, which exerts a force Ee on each particle (where E is the electric field and e is the ion or electron charge). However, electrons are much more mobile than ions, so that the latter can be considered as essentially immobile. As a charged particle travels along a field, it acquires a kinetic energy equal to its charge times the variation of potential. Thus, as electrons move downward in the leader, they accelerate and become able to ionize by collisions a number of other molecules or atoms. With each of the new electrons produced the process is repeated, with the result that the number of electrons increases exponentially: this is an avalanche. The increasing number of positive ions remaining behind creates a positive charge that opposes the pre-existing propelling field and prevents the avalanche from growing indefinitely.

As the leader progresses, the ionized conductive channel left behind brings downwards the high negative potential of the cloud negative centre: the tip is effectively connected to the starting point in the cloud. This is why, when approaching the ground, the field becomes very high between leader tip and ground. As the connecting discharge starts from below and propagates upward to meet the leader tip, it similarly produces an ionized channel at ground potential. Thus there is an extremely high difference of potential between the two conducting channels that meet to initiate the return stroke. The consequence is the extremely high current occurring in this stroke, first at the junction and then propagating upwards. This current is carried by a massive displacement of electrons downwards[*]. Intense luminosity, photoionization and heat result from this violent process.

A last remark may be made about the enormous energies involved. These do not result from large charges but from the large potential differences, as is characteristic of electrostatic situations in contrast to the more familiar domestic or industrial applications of electricity, where rather large amounts of charge flow through modest differences of potential. Thus a whole flash may discharge 30 C, a charge that flows through a 100 W light bulb in about a minute. But the difference of potential involved (that between ground and the negative space centre of the cloud) may be around 10^8 V. Therefore the electrostatic energy released during a flash is of the order of

$$30\,C \times 10^8\,V = 3 \times 10^9\,J$$

(equivalent, for instance, to the work necessary to lift $\sim$ 3000 tons 100 m above ground, or to the energy spent by a light bulb during almost a year).

Representative values of the main parameters related to lightning are summarized in Table 1.

TABLE 1

Typical values for cloud-to-ground lightning discharge bringing negative charge to Earth.

Stepped Leader	
Length of step	50 m
Time interval between steps	50 μs
Average velocity of propagation of stepped leader	1.5×10^5 m/s
Charge deposited on stepped-leader channel	5 C
Dart Leader	
Velocity of propagation	2.0×10^6 m/s
Charge deposited on dart-leader channel	1 C
Return Stroke	
Velocity of propagation	5.0×10^7 m/s
Peak current	10–20 kA
Charge transferred excluding continuing current	2.5 C
Channel length	5 km
Continuing Current	
Peak current	200 A
Time duration	0.1–0.2 s
Charge transferred	15 C
Lightning Flash	
Number of strokes per flash	3–4
Time interval between strokes in absence of continuing current	40 ms
Time duration of flash	0.2 s
Charge transferred including continuing current	25 C

[*] Notice that this does not mean that each electron travels all the way from cloud to ground, but rather that all the electrons along the channel are displaced downwards with respect to the positive ions, thereby producing an effective transfer of negative charge from cloud to ground. Thus during the stroke, Coulombs of charge are transferred across several kilometres, while each electron has travelled a distance of the order of metres.

Chapter VI: Questions

Q1. Consider the slopes of the curves in Figure 3 and Figure 5. What does the value of each one of them measure?

Q2. How do you explain that for increasing altitude, the conductivity of the air increases, but its space charge density decreases?

Q3. Knowing that cosmic rays consist of protons and other charged particles, would you expect them to penetrate uniformly through the atmosphere, or would you expect a latitude effect affecting their contribution to ion production? Explain.

Q4. What is a *small ion*? A *large ion*?

Q5. What molecular parameters do you think are important in determining the possibility of these two ion-molecule charge exchange reactions:

$$N_2^+ + O_2 = N_2 + O_2^+$$
$$O_2^- + NO_2 = O_2 + NO_2^- \quad ?$$

Q6. Why do you think that molecular ions cluster with neutral molecules, while neutral molecules like N_2, O_2, etc. remain single?

Q7. From the fact that conductivity increases with height in the troposphere, can you conclude that the downward electric field must decrease? Explain.

Q8. In what conditions of purity or pollution does the concentration of small ions in the air become approximately inversely proportional to the number concentration of aerosol particles?

Q9. If the positive space charge in the atmosphere moves towards the ground under the influence of the electric field in fair weather, why does it not disappear?

Q10. Assume that an isolated thundercloud with the usual polarity passes over an observation station. What changes will be observed in the vertical component of the electric field at ground?

Q11. Why does lightning not occur in shallow layer clouds?

Q12. Does cloud-to-ground lightning contribute to the charging between Earth and ionosphere? And in-cloud lightning between the main charge centres?

Q13. Why is the return stroke much faster and much more intense than the stepped leader?

Q14. Why does a connecting discharge start from a point on the ground only when the stepped or dart leader arrives close to the ground? Explain.

Chapter VI: Problems

(Any necessary constants not given in the statement of a problem will be found in the Table of Constants on pages x–xi)

P1. A common instrument used to measure the electric field in the atmosphere is the 'field mill'. In it, insulated horizontal plates are covered and uncovered successively by rotating vanes, electrically grounded. From a connection to the insulated plates, an alternating signal is obtained, which can be amplified and used to measure the field. Explain how this signal is produced.

P2. Assume that in the first 200 m, the space charge ρ can be expressed in elementary charges per cm^3 by

$$\rho = 20.4 \, e^{-0.00452 \cdot z}$$

where the altitude z is in m.
    (a) Obtain expressions for $E(z)$ and $V(z)$ (where E = electric field and V = electric potential) valid between 0 and 200 m.
    (b) Give ρ, E and V at 200 m over the ground.
$E_0 = -120 \, \text{V/m}$ and $V_0 = 0$.

P3. Assume that there are 1000 positive ions and 900 negative ions per cm^3 in the air. Calculate the space charge density, in $p\text{C/m}^3$.

P4. Assume that a wire mesh spherical cage of 1 m radius is exposed to the atmosphere and electrically grounded. A device is used to measure the potential V at the centre of the cage, and gives the value 0.37 V (with respect to ground potential, taken as 0). What is the charge density in the air, assumed uniform? Take into account that, according to Coulomb's law, the field of a spherically-symmetrical distribution of charge is

$$E = \frac{1}{4\pi\epsilon_0} \frac{Q}{r^2}$$

for any r, where E is the radially directed field, ϵ_0 = permittivity of free space, Q = total charge within the radius r, r = distance from the centre of the charge distribution.

P5. Assume that small ions are being produced at the rate of $q = 10 \, \text{pairs/cm}^3.\text{s}$, that the recombination coefficient is $\alpha = 1.6 \times 10^{-6} \, \text{cm}^3/\text{s}$, the attachment coefficient is $\beta = 10^{-6} \, \text{cm}^3/\text{s}$ and that the number concentration of particles is $N = 10^4 \, \text{cm}^{-3}$.
    (a) What is the equilibrium number concentration of ions n (ignore differences between positive and negative ions)?

(b) What are the fractions that disappear by recombination and by attachment?

(c) Repeat the calculations for $N = 100 \text{ cm}^{-3}$.

P6. The mobility k of an ion is defined by the relation

$$\mathbf{v} = k\mathbf{E}$$

where $\mathbf{v}$ is the drift velocity the ion acquires under the influence of an electric field $\mathbf{E}$. Assuming $k \simeq 1.5 \text{ cm}^2/\text{V.s}$ for all ions, and assuming that the concentration of positive and negative ions is 1000 cm^{-3} of each sign, calculate the conductivity λ of the air.

P7. A metal sphere with a charge Q_0 is exposed to the air on an insulating support. The air has a conductivity of $2 \times 10^{-14} \Omega^{-1} \text{m}^{-1}$, to which positive and negative ions are assumed to contribute equally, and the air flow maintains this conductivity close to the sphere surface.

(a) Find an expression for the decay of the charge Q_0.

(b) Find the *relaxation time* for this decay, i.e. the time at which Q has decayed to e^{-1} of its original value Q_0.

The field of a charge distribution of spherical symmetry is given by

$$E = \frac{1}{4\pi\epsilon_0} \frac{Q}{r^2}$$

where r is the distance from the centre and ϵ_0 is the permittivity of free space.

P8. Assume that in a thundercloud, the main charges are $+ 50 \text{ C}$ aloft and $- 50 \text{ C}$ at lower levels. Compute the electric field at the middle point between the charges assuming two approximate models:

(a) The charges are extended horizontally over a radius of 5 km, and separated by only 2 km, so that the system can be approximately treated as a parallel plate capacitor.

(b) Each of the charges is distributed with spherical symmetry, and the centres are separated by 2 km.

The capacity of a parallel plate capacitor is $C = \epsilon_0 A/d$, where ϵ_0 is the permittivity of free space, A the area of the plates and d the distance between them. The field of a spherically symmetrical distribution of charges is

$$E = \frac{1}{4\pi\epsilon_0} \frac{Q}{r^2}$$

where Q is the total charge and r the distance from the centre (for points outside the charged region).

P9. (a) What is the power developed at a peak current of $10\,000 \text{ A}$ during the return stroke in a cloud-to-ground lightning, if the potential difference between cloud and ground is 10^8 V?

(b) What is the average power for a total duration of the lightning of 0.2 s, if this transferred a total of 20 C?

VII. Atmospheric Dynamics

Atmospheric dynamics is the study of air motion. This motion, as that of any object in classical physics, obeys Newton's laws. One needs to understand first the nature of the forces experienced by the air before one can properly understand the complex ways through which the air moves in the atmosphere. The air in the Earth's atmosphere experiences basically four types of forces, known as the gravitational force, the pressure gradient force, the Coriolis force, and the frictional force.

The motion of air has a very wide spectrum, both in spatial and in temporal scales, ranging from the random motions of the molecules to the global circulation involving the entire atmosphere. Except for that part of air near the outer edge of the Earth's atmosphere, the motions of individual molecules are not very important to the understanding of atmospheric dynamics. It is usually adequate to consider the air as a continuum of fluid. When considered as such, the atmosphere has infinite degrees of freedom. We must find some proper ways to describe the velocity field in the atmosphere.

1. The Description of Air Motion

It is possible to describe the motion of the air in two distinct ways. One is called the Eulerian description and the other the Lagrangian description.

1.1. *The Eulerian description*

The Eulerian description consists in specifying the air properties as functions of position in space (r) and in time (t). The primary quantity of air motion is the velocity V which is considered as a function of position and time $V(r, t)$. This Eulerian description can be thought of as providing a picture of the spatial distribution of air velocity at each instant during the motion. Obviously, if we fix our attention to a given element of volume, the air that fills that element is changing continuously.

1.2. *The Lagrangian description*

In the Lagrangian description, the air is considered as a collection of small air parcels. The velocity of each air parcel is described as a function of time. It gives the dynamical history of each and every air parcel of the atmosphere. At any given instant, the Eulerian velocity field may be constructed from the Lagrangian description by knowing the positions of air parcels, and their velocities at that instant.

The Lagrangian description, although intuitively attractive, often leads to very cumbersome analysis. An air parcel often changes its shape when it moves. The Lagrangian description usually does not give directly the spatial gradients of velocity which are needed to determine interactions between air parcels. We shall, therefore, employ primarily the Eulerian description in the following pages.

1.3. *Basic types of velocity distribution within a small air parcel*

A velocity field within a small air parcel, no matter how complex it may appear, can always be considered as a superposition of a few simple types of motion field. The first basic type of velocity distribution, called *linear translation*, is best illustrated in Figure 1(a) where the velocity is a constant within the air parcel. The air parcel moves linearly with a constant speed without changing its shape.

The second type of velocity distribution, called *rotation*, is illustrated in Figure 1(b), where the air parcel rotates like a solid rigid body.

Sometimes the velocity distribution within an air parcel may cause the parcel to expand (or to contract) without changing its shape. This type of motion field is called *isotropic expansion* and is illustrated in Figure 1(c). The fourth type of motion field has a velocity distribution which will cause the air parcel to change its shape without changing its volume. It is illustrated in Figure 1(d), where the spherical fluid element is being deformed into an ellipsoid. Motion fields of the types shown in Figures 1(c) and 1(d) are also called *pure straining motions*.

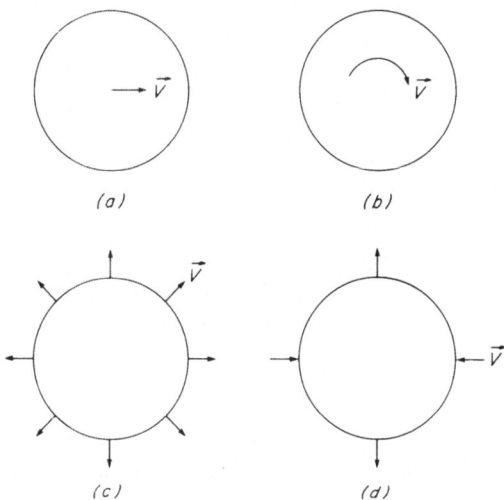

Fig. VII-1. *Basic types of air motion in a small air parcel.* The velocity field can be considered as the superposition of four basic types of motion: (a) linear translation, (b) rigid-body rotation, (c) isotropic expansion, and (d) pure straining motion without changing the volume of the parcel.

To illustrate how a velocity field may be decomposed into the four basic types of motions shown in Figure 1, let us consider the simple velocity distribution given in Figure 2(a). The velocity is everywhere parallel to the x-axis and it varies linearly in the y-direction. This kind of motion is known as the *simple shearing motion*.

Consider the point a on the surface of a cylindrical fluid element shown in Figure 2(b). The velocity $\mathbf{V}$ at this point can be considered as the vector sum of the two vectors $\mathbf{R}_a$ and $\mathbf{S}_a$, with $\mathbf{S}_a$ normal to the surface and $\mathbf{R}_a$ tangential to the surface of the fluid element. Similar decompositions can be made of the velocities at points b, c and d. The velocity vectors $\mathbf{R}_a$, $\mathbf{R}_b$, $\mathbf{R}_c$ and $\mathbf{R}_d$ represent a rigid body rotation of the fluid element. The remaining components, $\mathbf{S}_a$, $\mathbf{S}_b$, $\mathbf{S}_c$ and $\mathbf{S}_d$ represent a pure straining motion without changing the volume of the element.

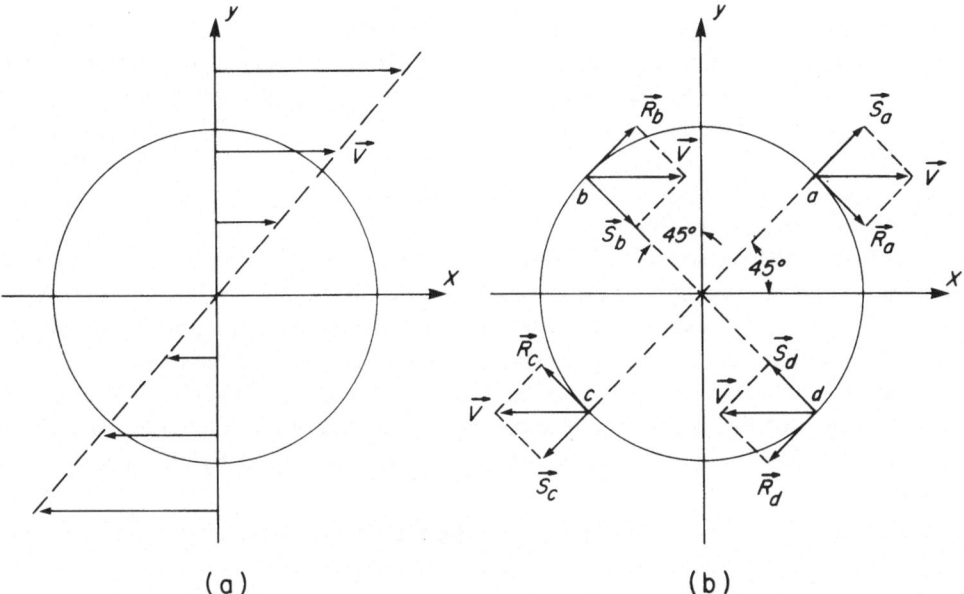

(a) (b)

Fig. VII-2. *Simple shearing motion.* A simple shearing motion is shown in (*a*). The velocity has the same direction everywhere and varies linearly in the direction perpendicular to the velocity vector. The simple shearing motion is decomposed in (*b*) into a rigid body rotation ($\mathbf{R}_a$, $\mathbf{R}_b$, $\mathbf{R}_c$ and $\mathbf{R}_d$) and a pure straining motion ($\mathbf{S}_a$, $\mathbf{S}_b$, $\mathbf{S}_c$ and $\mathbf{S}_d$).

2. The Principal Forces Acting on a Parcel of Air in the Atmosphere

As already mentioned, the motion of a material body obeys Newton's laws of motion. A material body moves with a constant velocity unless it is acted upon by a force or forces. To fully understand the motion of air in the atmosphere, one must first understand the nature of the forces experienced by an air parcel.

To an observer who is stationary with respect to the Earth's surface, there are four major forces acting on an air parcel in the atmosphere: the gravitational force, the pressure gradient force, the Coriolis force and the frictional force. The nature of these forces will be discussed briefly.

2.1. *The gravitational force*

Due to the large mass of the Earth, the gravitational force is one of the strongest forces experienced by an air parcel. The gravitational force is directed toward the centre of mass of the Earth.

If $\mathbf{g}$ is the gravitational acceleration, the gravitational force experienced by a small air parcel of volume Δv is given by

$$\Delta v \rho \mathbf{g}$$

where ρ is the density of the air.

Because of the rotation of the Earth around its own axis, an air parcel fixed with

151

respect to the Earth's surface also experiences a centrifugal force directed away from the axis of rotation. The centrifugal force is very small in comparison with the gravitational force. It is often combined with the gravity to form the so-called *effective gravity*. The difference between the effective gravity and the gravity is very small and can usually be neglected.

2.2. *The pressure gradient force*

The pressure at the surface of an air parcel is the normal component of the force exerted by its surroundings on a unit area of the surface. This force is always directed toward the parcel. The parcel will experience a net force only if there is a difference between the pressures on surfaces at opposite sides. This is called the pressure gradient force.

Consider a column of air as shown in Figure 3. ΔA is the cross-sectional area and δx is the length of the air column. The pressure at one end of the column is p and at the other end $p + \delta p$. The total force in the x-direction experienced by the air column is given by

$$F_x = p\Delta A - (p + \delta p)\Delta A = -\delta p \Delta A \tag{1}$$

If ρ is the density of the air, the force per unit mass is then given by

$$\frac{F_x}{\rho \Delta A \delta x} = -\frac{1}{\rho}\frac{\delta p \Delta A}{\delta x \Delta A} \equiv -\frac{1}{\rho}\frac{\partial p}{\partial x} \tag{2}$$

This result can be easily generalized. If p is the pressure field of the air, the force per unit mass on an air parcel is given by

$$\mathbf{f} = -\frac{1}{\rho}\nabla p \tag{3}$$

The gravitational force and the pressure gradient force are the only two forces that can initiate motion of the air.

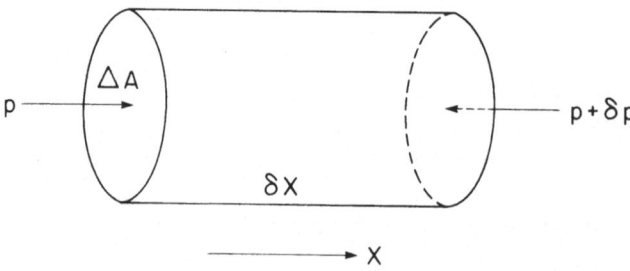

Fig. VII-3. *Pressure gradient force.* When the pressures at opposite ends of an air parcel are different, the parcel experiences a net force. The force per unit volume is $- \nabla p$. It is in the direction opposite to the gradient of pressure.

2.3. *The Coriolis force*

Because of the rotation of the Earth around its own axis, a coordinate system fixed to the Earth's surface is not an inertial system. A force must be introduced in the equation of motion written in this coordinate system to reflect this fact. The force is called the Coriolis force.

In order to understand this, let us start with a simple example. Consider a circular region around the North Pole. If it is small enough, we can think of it as a rotating disc, with the angular velocity Ω of the Earth's rotation. Now let us consider a body, in our case an air parcel, starting to move horizontally away from the Pole P in the direction towards point A, as indicated in Figure 4. If no force acts on this parcel, it will follow a uniform motion with respect to an inertial frame (i.e. to fixed stars: Newton's law). But, as the disc rotates, the projection of the motion over its surface follows a curved line such as PA', rather than the straight line PA. In fact, the motion has been straight, but by the time the parcel arrives at the periphery of the disc, A' has come to the original position of A, while A (as a point fixed to the surface) is now in the position B.

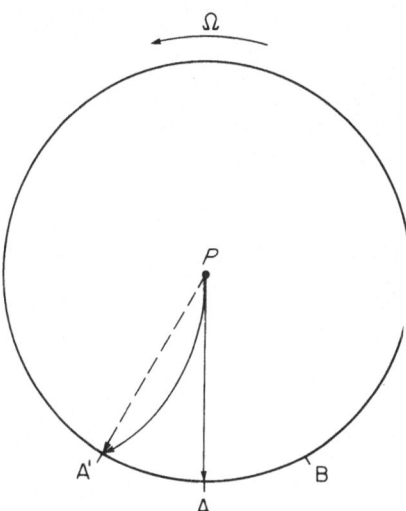

Fig. VII-4. *The Coriolis force.* A particle moving along a straight line PA will trace out a curve trajectory PA' on a rotating disk, with an angular velocity Ω. To an observer stationary with respect to the disk, the curved path reflects the presence of a force. It is known as the Coriolis force.

To an observer rotating with the Earth, the curved line PA' will appear as the trajectory; if he overlooks the fact that the Earth rotates (and himself with it) with respect to an inertial frame, the trajectory will look to him as if a deflecting force towards the right (looking from P) was acting on the parcel. This fictitious force is the Coriolis force.

Let us now consider the motion of an air parcel at any latitude ϕ over the surface of the Earth. Let u be the component of the air parcel velocity tangential to the latitudinal circle, v be the component going away from the axis of rotation of the Earth, as shown in Figure 5. Ω is the angular velocity of the Earth and R is the Earth's radius. Let S be used

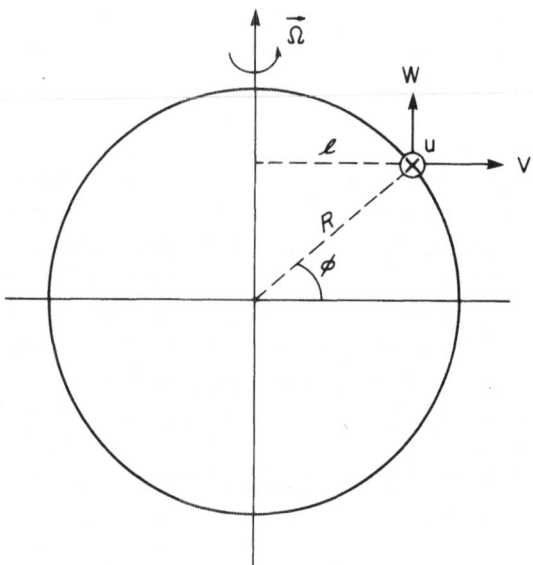

Fig. VII-5. *Motion of an air parcel.* u is the eastward component, v the component directed away from the Earth's axis, and w the component parallel to the Earth's axis $\ell = R \cos \phi$. R is the radius of the Earth and ϕ is the latitude. Ω is the angular velocity of the Earth's rotation.

to denote a coordinate system fixed with respect to the Sun and E to denote a coordinate system fixed to the Earth.

The centrifugal acceleration experienced by the air parcel is given by

$$a_c = u^2/\ell \qquad \ell = R \cos \phi \tag{4}$$

in the E system, and by

$$a_0 = \frac{(\Omega \ell + u)^2}{\ell} = \Omega^2 \ell + 2\Omega u + \frac{u^2}{\ell} \tag{5}$$

in the S system. The two additional terms $\Omega^2 \ell$ and $2\Omega u$ in the inertial system S would then appear to an observer in the E system as a force per unit mass. The term $\Omega^2 \ell$ is independent of the air parcel velocity and hence can be combined with the gravitational acceleration to give an effective gravity. The other term, $2\Omega u$, directed away from the Earth's axis, is one component of the Coriolis force.

The other component of the Coriolis force may be obtained by considering the conservation of angular momentum.

The rate of change of angular momentum can be written as

$$\frac{d}{dt} \ell u = \ell \frac{du}{dt} + u \frac{d\ell}{dt} = \ell \frac{du}{dt} + uv \tag{6}$$

in the E coordinate system and as

$$\frac{d}{dt} \ell(\Omega \ell + u) = 2\Omega \ell v + \ell \frac{du}{dt} + uv \tag{7}$$

154

in the S system. The additional acceleration term $2\Omega v$ in the direction opposite to u in the inertial system S can be interpreted as a force in the E system. This term is the other component of the Coriolis force. Combining the two components, one finds that the Coriolis force per unit mass that has to be introduced in the E system can be written as

$$\text{Coriolis acceleration} = -2\,\boldsymbol{\Omega} \times \mathbf{V} \tag{8}$$

2.4. *The frictional force*
There exists considerable frictional force between the atmosphere and the Earth's surface. For instance, it is clear that the ground has a braking effect on the first layers of moving air. Irregularities on the ground surface sometimes make this more apparent; it is well known that lines of tall trees are planted to 'break' undesirable winds and decrease their damaging effect on certain crops. Frictional force may also exist between an air parcel and its surroundings.

The nature of this frictional force is a rather difficult subject. Molecular viscosity is only important within the lowest metre or so above the Earth's surface. Most of the frictional force in the atmosphere is produced by smaller scale eddy mixing processes.

3. The Acceleration of an Air Parcel

According to Newton's Second Law of Motion, the acceleration of an air parcel is equal to the total force exerted on the parcel divided by its mass. We will first demonstrate in this section how the acceleration of an air parcel may be determined from the Eulerian description of the velocity field in the atmosphere. The equation of motion that can be used to determine the Eulerian velocity field will then be introduced.

3.1. *The total derivative*
In the Eulerian description, the velocity field $\mathbf{V}(\mathbf{r}, t)$ is defined as a function of position $\mathbf{r}$ and time t. $\mathbf{r}$ is used here to denote the position vector

$$\mathbf{r} = \hat{\mathbf{x}}x + \hat{\mathbf{y}}y + \hat{\mathbf{z}}z$$

$\hat{\mathbf{x}}$, $\hat{\mathbf{y}}$ and $\hat{\mathbf{z}}$ are the three unit vectors in a Cartesian coordinate system. For an air parcel located at position $\mathbf{r}$ at time t, the acceleration of the parcel has two distinctly different components. The first component is due to the rate of change of the fluid velocity at point $\mathbf{r}$. It is given by the partial derivative of $\mathbf{V}$ with respect to t:

$$\frac{\partial \mathbf{V}(\mathbf{r}, t)}{\partial t}$$

This component is called the *local derivative*. The second component is due to motion of the air parcel. The position of the parcel is a function of time. The parcel will experience an acceleration as it moves through the velocity field even if the velocity field is not changing in time. This component is called the *advective derivative*. It can be easily understood by considering a stationary, one-dimensional velocity field. We shall assume that the

155

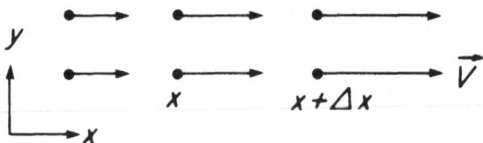

Fig. VII-6. *Acceleration of an air parcel in a stationary velocity field.* A stationary velocity field is shown with the velocity **V** parallel everywhere to the *x*-axis and the speed increasing with *x*. An air parcel moves along such a velocity field and experiences an acceleration, even though the local time derivative of **V** is zero everywhere.

velocity of the fluid is in the x direction and the speed V increases with x as shown in Figure 6. Suppose the positions of an air parcel at times t and $t + \Delta t$ are x and $x + \Delta x$ respectively. The acceleration of the parcel at time t is given by

$$\lim_{\Delta t \to 0} \frac{V(x + \Delta x) - V(x)}{\Delta t} = \lim_{\Delta t \to 0} \frac{\Delta x}{\Delta t} \frac{V(x + \Delta x) - V(x)}{\Delta x}$$

The limit of $\Delta x/\Delta t$ as Δt goes to zero gives the rate of change of position of the parcel which is the velocity of the parcel at time t. The acceleration is therefore given by

$$V \frac{\partial V(x)}{\partial x}$$

Thus, due to the fact that the parcel is moving across a velocity field which is not uniform in space, it experiences an acceleration even if the field is independent of time.

These considerations can be easily extended to a general Eulerian velocity field $\mathbf{V}(\mathbf{r}, t)$. Let us consider first the x-component of the velocity $V_x(\mathbf{r}, t)$. Assume that the positions of an air parcel at t and $t + \Delta t$ are $\mathbf{r}$ and $\mathbf{r} + \Delta \mathbf{r}$. The x-component of the acceleration, a_x, is

$$a_x = \lim_{\Delta t \to 0} \frac{V_x(\mathbf{r} + \Delta \mathbf{r}, t + \Delta t) - V_x(\mathbf{r}, t)}{\Delta t}$$

$$= \lim_{\Delta t \to 0} \left(\frac{\partial V_x}{\partial t} \Delta t + \frac{\partial V_x}{\partial x} \Delta x + \frac{\partial V_x}{\partial y} \Delta y + \frac{\partial V_x}{\partial z} \Delta z \right) \bigg/ \Delta t \qquad (9)$$

The limits of $\Delta x/\Delta t$, $\Delta y/\Delta t$ and $\Delta z/\Delta t$, give the three components, V_x, V_y and V_z, of the velocity of the air parcel at time t. The acceleration a_x is thus given by

$$a_x = \frac{\partial V_x}{\partial t} + V_x \frac{\partial V_x}{\partial x} + V_y \frac{\partial V_x}{\partial y} + V_z \frac{\partial V_x}{\partial z} \qquad (10)$$

It is often expressed in a more compact form:

$$a_x = \frac{\partial V_x}{\partial t} + (\mathbf{V} \cdot \nabla) V_x \qquad (11)$$

The notation $\mathbf{V} \cdot \nabla$ is used to denote the operator

$$\mathbf{V} \cdot \nabla = V_x \frac{\partial}{\partial x} + V_y \frac{\partial}{\partial y} + V_z \frac{\partial}{\partial z} \qquad (12)$$

Formally, it can be interpreted as the dot product of the vector

$$\mathbf{V} = \hat{x}V_x + \hat{y}V_y + \hat{z}V_z \tag{13}$$

and the gradient operator

$$\nabla = \hat{x}\frac{\partial}{\partial x} + \hat{y}\frac{\partial}{\partial y} + \hat{z}\frac{\partial}{\partial z} \tag{14}$$

Similar calculations show that a_y and a_z, the accelerations in the $\hat{y}$ and $\hat{z}$ directions are given by

$$a_y = \frac{\partial V_y}{\partial t} + (\mathbf{V} \cdot \nabla)\, V_y \tag{15}$$

$$a_z = \frac{\partial V_z}{\partial t} + (\mathbf{V} \cdot \nabla)\, V_z \tag{16}$$

The three components may be combined using vector notation, to write

$$\mathbf{a} = \frac{\partial V}{\partial t} + (\mathbf{V} \cdot \nabla)\,\mathbf{V} \tag{17}$$

The total rate of change of the velocity of an air parcel has thus two components. The first part is the local time derivative and the second part the advective time derivative. The two components, when combined, are called the *total time derivative*. The total derivative of the Eulerian velocity field gives the acceleration of an air parcel. It is usually written as

$$\frac{D\mathbf{V}}{Dt} = \frac{\partial \mathbf{V}}{\partial t} + (\mathbf{V} \cdot \nabla)\,\mathbf{V} \tag{18}$$

The notation D/Dt is used to denote the total derivative operation

$$\frac{D}{Dt} = \frac{\partial}{\partial t} + (\mathbf{V} \cdot \nabla) \tag{19}$$

It can be used to calculate the time rate of change of any properties of an air parcel described in the Eulerian framework.

3.2. *The equation of motion*

For an air parcel of unit volume, the air density times the acceleration should be equal to the vector sum of the forces acting upon the parcel. This equality is called the *equation of motion*. The expression of the acceleration of an air parcel has been derived in the previous section. It is given by the total time derivative of the velocity field. The equation of motion can be written as

$$\rho\frac{D\mathbf{V}}{Dt} = \rho\mathbf{g} - \nabla p - \rho 2\,\mathbf{\Omega} \times \mathbf{V} + \text{Frictional force} \tag{20}$$

where ρ is the air density, $\mathbf{g}$ is the effective gravitational acceleration, p the pressure, $\mathbf{\Omega}$ the angular velocity of the Earth and $\mathbf{V}$ the air velocity.

4. The Continuity Equation

The equation of motion derived in the previous section gives an equation which can be used to solve for the Eulerian velocity field in the atmosphere. The equation alone, however, is not complete in the sense that it requires the density and pressure distributions of the atmosphere before it can be solved. The density and pressure distributions, in turn, may be dependent — among other things — upon the velocity field of the air. Additional equations are obviously needed before the motions of the atmosphere can be determined as solutions of the equation of motion.

One of these additional equations is provided by the law of mass conservation. This law often imposes a very strong constraint on the flow patterns that can be realized physically in the atmosphere.

Consider a fixed, but otherwise arbitrary volume V in the atmosphere. The surface which encloses V will be denoted by S. The total mass that is enclosed by the surface S at any instant is given by

$$\int_V \rho \, dV$$

Because of air motion, air flows into or out of the volume V at various parts of the surface S. As a result, the total mass enclosed by surface S should be a function of time. The rate of increase of the total mass contained in V is given by

$$\frac{d}{dt} \int_V \rho \, dV$$

which should be equal to the net rate of inflow of air into the volume across the surface S.

The mass flux associated with the motion of air is given by the density of air ρ multiplied by its velocity $\mathbf{V}$. At a small element ΔS of surface S, only the component of the mass flux normal to the surface will produce any flow of mass out of the volume V, as shown in Figure 7. The unit vector which is normal to the surface element will be denoted by $\hat{n}$. We will adopt the convention that $\hat{n}$ is positive when it points outward from the volume. The rate of mass flow out of the volume through ΔS is thus given by $\rho \mathbf{V} \cdot \hat{n} \Delta S$. When integrated over the entire surface S, it gives the total rate of mass flow out of V:

$$\text{Total Rate of Mass Flow Out of } V = \int_S \rho \mathbf{V} \cdot \hat{n} \, dS \tag{21}$$

The law of mass conservation then requires that

$$\frac{d}{dt} \int_V \rho \, dV = -\int_S \rho \mathbf{V} \cdot \hat{n} \, dS \tag{22}$$

Since the volume V is fixed, the time derivative may be carried out under the integral sign to give

$$\int_V \frac{\partial \rho}{\partial t} \, dV + \int_S \rho \mathbf{V} \cdot \hat{n} \, dS = 0 \tag{23}$$

This integral form, or the global statement of mass conservation may be transformed into a differential form, or a local statement by applying the *Gauss theorem* in vector analysis.

158

Fig. VII-7. *Air flow out of a volume through the bounding surface* S. The rate of air flow out of a volume through a small surface element ΔS is $\rho \mathbf{V} \cdot \hat{n} \Delta S$, where ρ = air density, $\mathbf{V}$ = air velocity, $\hat{n}$ = unit vector normal to the surface element ΔS. The total rate of outflow is given by the summation of this quantity over the entire bounding surface S.

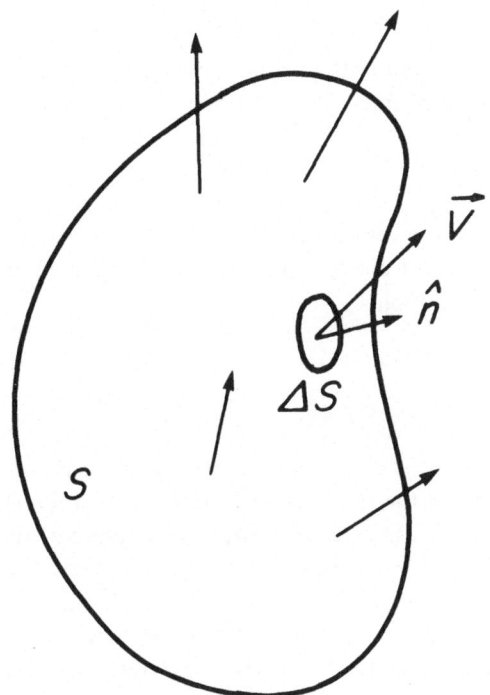

The Gauss theorem states that for a reasonably smooth vector field $\mathbf{F}$, the surface integral of $\mathbf{F}$ over S is equal to the volume integral of the *divergence* of $\mathbf{F}$, $\nabla \cdot \mathbf{F}$, over the volume V enclosed by S:

$$\int_S \mathbf{F} \cdot \hat{n} \, dS = \int_V \nabla \cdot \mathbf{F} \, dV \tag{24}$$

The proof of the theorem can be found in any standard book on vector analysis.

By applying the Gauss theorem, Equation (23) can be transformed into

$$\int_V \left(\frac{\partial \rho}{\partial t} + \nabla \cdot \rho \mathbf{V} \right) dV = 0 \tag{25}$$

This equation is valid for any volume V in the atmosphere, which is possible only if the integrand itself is zero everywhere. Hence

$$\frac{\partial \rho}{\partial t} + \nabla \cdot \rho \mathbf{V} = 0 \tag{26}$$

The equation is a local representation of the law of mass conservation. It is called the *continuity equation*.

The continuity equation can be written in a different form, using the notation of the total derivative:

$$\frac{1}{\rho}\frac{D\rho}{Dt} + \nabla \cdot \mathbf{V} = 0 \tag{27}$$

In certain circumstances, the temporal and spatial variations of ρ are small and the total derivative of air density becomes negligible. The continuity equation then takes the simple form

$$\nabla \cdot \mathbf{V} = 0 \tag{28}$$

In this case, the three components of the velocity are not independent from one another. If two components of the velocity vector are known, the third component can be determined without using the equation of motion.

5. The Scales of Motion

Generally speaking, the motion of the atmosphere is governed by the equation of motion, the continuity equation, and the laws of thermodynamics discussed in Ch.IV. An observed circulation in the atmosphere can be considered as a particular solution of the governing equations. These equations are all nonlinear partial differential equations. Their solution under a given set of physical and boundary conditions often cannot be easily obtained. In order to understand the physical and dynamical nature of atmospheric circulations, it is frequently necessary to first classify, according to a certain criterion, the various types of circulation systems observed in the atmosphere. A particular type of circulation can then be isolated and its dynamical nature be examined with the aid of the dynamic equations. With the better understanding so obtained of the nature of the circulation systems, certain approximations can then be introduced, so as to make the equations easier to solve.

The motion systems which occur in the atmosphere may be classified in various ways. One method of classification that has proved to be very useful is based on the time and distance scales by which the motion systems may be recognized. Atmospheric motions are often composed of a spectrum of circulation systems of different time and length scales. The time scale is usually related to its length scale; the larger the length scale, the longer its time scale. The largest circulation systems of the atmosphere have a length scale comparable to the diameter of the Earth itself. The smallest circulations have a length scale comparable to the mean free path of the individual molecules.

According to the length scale, the spectrum of atmospheric motions can be divided into *planetary-scale* motions, *synoptic-scale* motions, *mesoscale* motions, and *small-scale* motions*. The boundaries between these subdivisions are not well defined, for the spectrum of atmospheric motions is continuous. Nevertheless, the motion systems in each subdivision have their own distinct dynamic features. Different approximations can be introduced into the dynamical equations for motion systems in each subdivision. The

---

* This classification follows essentially that given in Wallace and Hobbs (see Bibliography). The very small end of the small-scale motions is sometimes known also as the *microscale*.

classification has been a very useful conceptual tool in the study of atmospheric dynamics.

The planetary-scale motions include circulation systems with a horizontal scale comparable to the dimension of the Earth. The mean global circulations to be discussed in §9, and the zonally varying features with length scale comparable to the major continents and oceans are examples of motion systems of this type.

Synoptic-scale motions have horizontal scales smaller than those of planetary-scale motions, yet large enough to be resolved by a conventional observation network. The spacing of stations in such a network is of the order of several hundred kilometres. Most circulation systems responsible for the day-to-day weather changes are systems of this category.

Motion systems with a horizontal scale of the order of 10^1–10^2 kilometers are called mesoscale motion systems. Mountain lee waves, thunderstorms, squall lines and hurricanes are a few examples of motion systems of this type.

Circulations with horizontal dimensions smaller than those of mesoscale motions are called small-scale motions. Small cumulus clouds, and convective and mechanical turbulent eddies near the surface of the Earth are typical examples. Motions systems of this type play a very important role in the dynamics of the lowest kilometre of atmosphere.

We can summarize this classification as shown in Table 1.

TABLE 1

Scales of atmospheric motions

Length Scale	Typical Dimensions (km)	Examples
Planetary	10 000	Hadley cell (§9.2)
Synoptic	1 000	Mid-latitude cyclones (§11)
Mesoscale	100	Thunderstorms
Small scale	≤ 10	Small cumulus

There are many other ways of classifying atmospheric motions. According to the degree of regularity, a flow field may be called either a laminar flow or a turbulent flow. Some air motions are induced mainly by the pressure gradient force, some others are induced by the buoyancy force. Air motions can also be generated by some instability mechanism. Some of the instability processes are mostly thermal in nature, like the vertical instabilities discussed in Ch.IV; some others are mainly mechanical in nature, such as the shear instability often observed very close to the Earth's surface. The discussions of atmospheric motion systems in this chapter will be based mainly on the classification according to length scales.

6. Some Important Features of Large-scale Atmospheric Motions

Planetary-scale and synoptic-scale motions of the atmosphere have a few important common features that are not shared by smaller scale motion systems. They will be loosely referred to hereafter as the large-scale systems. In this section we will discuss two of these features, namely, the hydrostatic equilibrium and the quasi-horizontal nature of the large-scale flow. Since the boundaries between the various scales of motions cannot be sharply defined, motions at the large end of mesoscale systems may also share some of these features.

6.1. *Hydrostatic equilibrium*

One important feature of large-scale motions is that hydrostatic balance is maintained.

The hydrostatic pressure at a point in the atmosphere is defined as the total weight of air column with a unit cross-sectional area above the point (cf. Ch.I). Or, in other words

$$p_{\text{hydrostatic}} = \int_z^\infty \rho g \, dz \tag{29}$$

The atmosphere is said to be in a hydrostatic equilibrium if the pressure of the air is equal to the hydrostatic pressure everywhere. On the average, the atmosphere is in a hydrostatic equilibrium state – assuming this equilibrium constitutes a very good approximation to most motion systems with a horizontal scale larger than a few hundred kilometres.

In a hydrostatic motion system, the vertical component of the equation of motion is reduced to

$$\frac{\partial p}{\partial z} - \rho g = 0 \tag{30}$$

The variation of the vertical component of air velocity cannot be determined from the vertical component of the equation of motion.

One interesting property of a hydrostatic atmosphere is that the total internal energy of the air becomes proportional to its total gravitational potential energy. The internal energy of an air column of unit cross-sectional area is given by

$$U = \int_0^\infty C_v T \rho \, dz \tag{31}$$

where C_v is the heat capacity of air at constant volume, referred to the unit mass. The total potential energy is given by

$$P = \int_0^\infty \rho g z \, dz \tag{32}$$

Using the relation of the hydrostatic pressure ($dp = -g\rho dz$), P may be expressed as

$$P = -\int_0^\infty z \, dp = -pz \big|_0^\infty + \int_0^\infty p \, dz \tag{33}$$

The product pz is zero at both $z = 0$ and $z = \infty$ (notice that when $z \to \infty$, p tends exponentially to 0),

$$P = \int_0^\infty p \, dz = \int_0^\infty R\rho T \, dz$$

$$= \frac{R}{C_v} U \tag{34}$$

162

where the ideal gas law has been used. Therefore, the internal energy and the potential energy of a hydrostatic air column are proportional to each other. Any change in the internal energy must be accompanied by a corresponding change in the potential energy. If the internal energy of an air column is increased by some heating process, the air column must expand vertically, thereby increasing its gravitational potential energy. The absorbed energy must then become distributed in such a way that the proportion (34) is maintained.

6.2. *The quasi-horizontal nature of large-scale flows*

Another important feature of large-scale flows in the atmosphere is that the vertical component of air velocity is much smaller than its horizontal components. The motion of air is almost parallel to the Earth's surface. This can be understood by the following considerations.

Consider a weather system with a typical horizontal scale L. The vertical scale Z of the system is usually limited by depth of the troposphere D.

$$Z \leqslant D \tag{35}$$

A parcel of air that travels through the weather system has a typical horizontal velocity u and vertical velocity w. The time it takes for the parcel to travel a horizontal distance L is (see Figure 8):

$$T \approx L/u \tag{36}$$

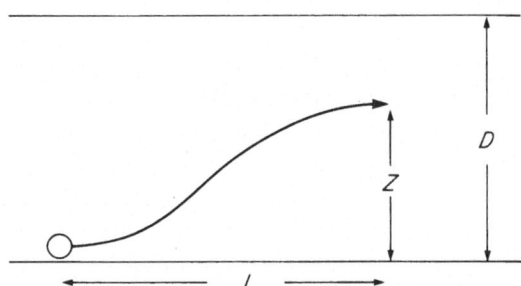

Fig. VII-8. *Trajectory of an air parcel moving through the troposphere.* An air parcel travels a horizontal distance L and a vertical distance z during a time interval Δt. Usually $z < D$, where D is the depth of the troposphere. The mean vertical velocity w of the parcel during the motion should have a magnitude such that $w/u \leqslant D/L$ where u is the mean horizontal velocity.

During the same time period, the parcel has also travelled a vertical distance $z \leqslant D$.

$$T \simeq z/w \leqslant D/w \tag{37}$$

Therefore,

$$\frac{L}{u} \leqslant D/w \tag{38}$$

Or

$$\frac{w}{u} \leqslant D/L \tag{39}$$

D is usually of the order of 10 km. If L is of the order of 10^3 km, w/u is at most of the order of 10^{-2}. Therefore, velocity vectors of large-scale motion systems are essentially two-dimensional in the horizontal plane. To determine the air motion, one needs to consider only the horizontal components of the equation of motion. This does not imply, however, that the vertical motion of the air is not important to the overall dynamics of the Earth's atmosphere.

If we let v_h be the horizontal velocity of an air parcel, the equation of motion for v_h can be obtained by taking the horizontal components of equation (20):

$$\rho \frac{Dv_h}{Dt} = -\nabla_h p - \rho f \hat{z} \times v_h + \text{Frictional Force} \tag{40}$$

where $\hat{z}$ is the unit vector in the vertical direction. $f = 2\Omega \sin \phi$ is called the Coriolis parameter. ϕ is the latitude.

7. The Relationship Between Pressure Gradient and Wind in Large-scale Middle Latitude Circulation Systems

The pressure gradient force is responsible for the initiation and maintenance of most large-scale atmospheric motions in the middle latitudes. As an example of the application of the equation of motion, we shall examine in this section the relationship between wind and pressure gradient force in these systems.

7.1. The geostrophic wind
In large-scale middle latitude systems, the accelerations of air parcels are usually small when compared with the Coriolis force exerted on them. The typical wind speed observed in the atmosphere is of the order of 10 m/sec. The Coriolis parameter in the mid-latitudes has a magnitude of $10^{-4} \, \text{s}^{-1}$. These give a Coriolis force of the order of $10^{-3} \, \text{m s}^{-2}$. The observed rate of change of wind velocity in the atmosphere (left side of (40)) is much smaller than this value. One may deduce that the Coriolis force must be approximately balanced by the pressure gradient force and the frictional force. In upper levels of the atmosphere where the frictional force is small, the Coriolis force must be balanced principally by the pressure gradient force. The air velocity determined by the balance of Coriolis and pressure gradient forces is called the *geostrophic wind*:

$$f \hat{z} \times v_g = -\frac{1}{\rho} \nabla p \tag{41}$$

where v_g is used to denote the geostrophic wind. For middle and high latitudes at least, the geostrophic wind is usually a good approximation to the observed wind velocity in the absence of frictional forces. Here, we shall demonstrate the relationship between the geostrophic wind and the pressure distribution.

The Coriolis force is perpendicular to the wind velocity. Its direction can be obtained by rotating the wind vector by $90°$ clockwise in the Northern hemisphere, as shown in Figure 9(a). In a geostrophic balance, the Coriolis force cancels the pressure gradient force exactly, as shown in Figure 9(b). Therefore the direction of the geostrophic wind can be obtained by rotating the vector of the pressure gradient force by $90°$ clockwise.

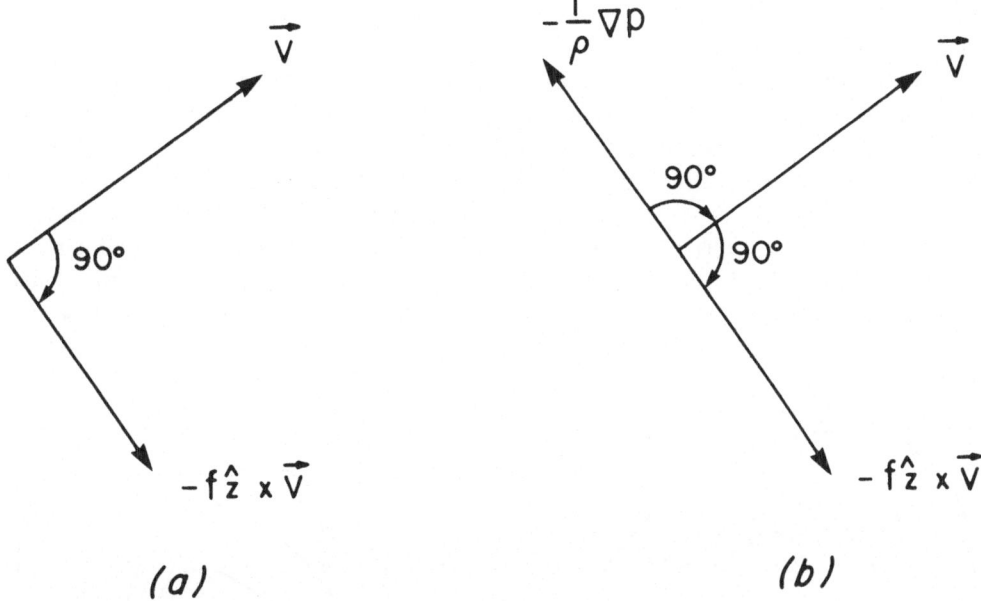

Fig. VII-9. *Geostrophic balance.* (*a*) The direction of the Coriolis force can be obtained in the Northern Hemisphere by rotating the wind vector by 90° clockwise. (*b*) The pressure gradient force is balanced by the Coriolis force in a geostrophic balance. The geostrophic wind direction can therefore be obtained by a 90° clockwise rotation of the pressure gradient force.

Hence, the geostrophic wind is parallel to the isobars. The wind vector should be in the direction with the low pressure to the left hand side of the wind vector. An example is given in Figure 10.

If the velocity of an air parcel is different from the geostrophic wind, the Coriolis force will not be able to balance the pressure gradient force. In the absence of frictional forces, the unbalanced forces will accelerate the air parcel. In that case, it may be shown that the air parcel will generally be accelerated in such a way that the geostrophic balance will be achieved.

From these discussions, it should become clear that over a low pressure area in the northern hemisphere, the geostrophic wind circulates in a counter-clockwise direction (clockwise in the southern hemisphere). This type of motion is called a *cyclonic motion.*

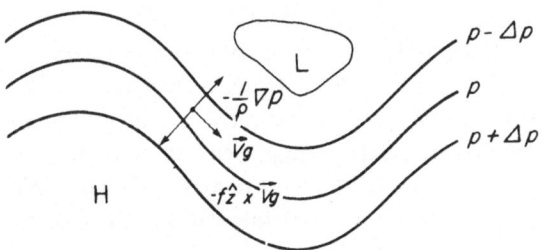

Fig. VII-10. *Geostrophic wind.* The geostrophic wind vector is parallel to the isobars with the low pressure region to the left-hand side of the wind vector.

165

A low pressure weather system is accordingly called a *cyclone*. An example of cyclonic motion is shown in Figure 11(a).

Over a high pressure area in the northern hemisphere, the geostrophic wind circulates in a clockwise direction (anticlockwise in the southern hemisphere). This type of motion is called *anticyclonic motion*. A high pressure system is often referred to as an *anticyclone*. An example of anticyclonic motion is shown in Figure 11(b).

Cyclonic systems can be found in both low and middle latitudes. Middle latitude cyclones are the systems which are responsible for the 'bad weather' in mid-latitude regions, with cloudy sky and precipitation as usual features. Because of their importance in weather forecasting, we shall discuss the detailed dynamic characteristics of these systems in §11.

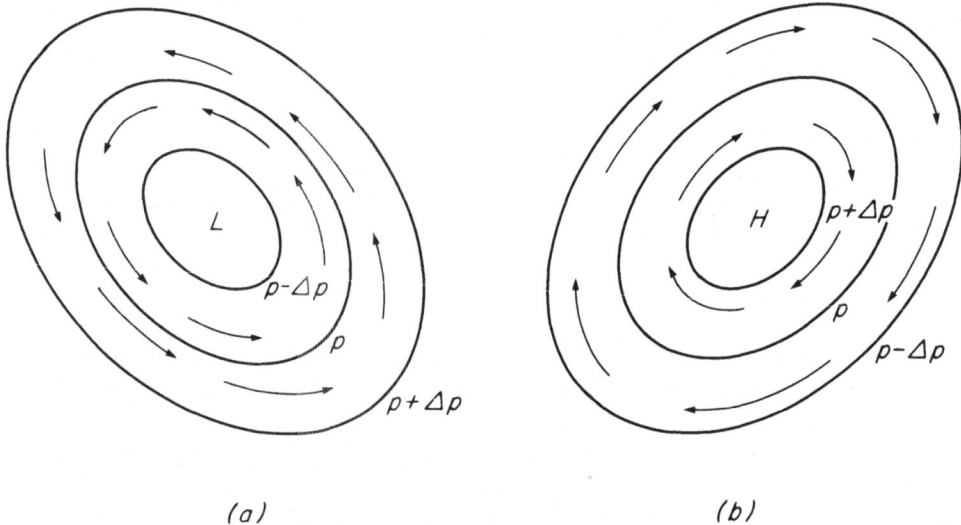

(a)   (b)

Fig. VII-11. *Cyclone and anticylcone*. (a) The geostrophic wind circulates around a low pressure centre in the counter-clockwise direction in the Northern Hemisphere. This type of motion is called cyclonic motion. A low pressure system is often referred to as a 'cyclone'. (b) A high pressure system is also called an 'anticyclone'. Geostrophic wind rotates in the Northern Hemisphere in the clockwise direction around the high pressure centre. Motions of this type are called anticyclonic motions.

Tropical cyclones are also known as *hurricanes* or *typhoons*. These are low pressure systems of extreme intensity. Wind speed in a tropical cyclone can be as high as 100 m/s. Extremely heavy rainfall is always produced. The cyclonic nature of air circulation in the lower troposphere in a hurricane or typhoon can be clearly seen from the pattern and movement of clouds from a satellite. An example is shown in Figure 12. It is a picture of clouds in Hurricane Ella, taken in the visible range of the radiation spectrum from NOAA5 satellite on 3 September 1978. Figure 13 gives another example, showing the Apollo 7 Earth-sky view of Hurricane Gladys, located at 150 miles southwest of Tampa, Florida.

Fig. VII-12. *Hurricane Ella*. A picture taken in the visible range of the radiation spectrum by NOAA5 weather satellite at 1410 Z, 3 September 1978. The white lumps are the tops of cumulonimbus clouds, while the more uniform fibrous layer consists of cirrus clouds. If the picture were a moving picture, it would show from the motions of clouds that air is spiralling cyclonically inward at lower levels and is spiralling anticyclonically outward at upper levels.

Fig. VII-13. *Hurricane Gladys*. Apollo 7 Earth-sky view of Hurricane Gladys, located at 150 miles southwest of Tampa, Florida. The slanting angle allows a view of the vertical structure of the hurricane.

7.2. *The variation of geostrophic wind with height*

For a layer of atmosphere in hydrostatic equilibrium, the pressure decreases with height. The 'thickness' of the layer of air between two isobaric surfaces is generally proportional to the mean temperature of air in the layer. This can be shown by the hydrostatic relationship (cf. Ch.I, §2):

$$dp = -\rho g \, dz \tag{42}$$

Consider two isobaric surfaces with pressures p_1 and p_2. Let z_1 and z_2 be the heights of the isobaric surfaces, respectively. We shall assume $z_2 > z_1$. It follows then that $p_1 > p_2$. Using the ideal gas law, the hydrostatic relation can be rearranged to give

$$-\frac{RT}{Mg}\frac{dp}{p} = dz \tag{43}$$

where R is the gas constant and M the average molecular weight of air. Vertical integration of the equation from z_1 to z_2 gives

$$\frac{R}{Mg}\int_{p_1}^{p_2} T \, d\ln p = z_2 - z_1 \tag{44}$$

where g has been taken as approximately constant. If we define $\bar{T}$ as the mean temperature of the layer according to the mean value theorem, the thickness of the layer is given by

$$z_2 - z_1 = \frac{R\bar{T}}{Mg}\ln\frac{p_1}{p_2} \tag{45}$$

Therefore, in a hydrostatic atmosphere, the warmer the air the larger is the distance between two isobaric surfaces. This fact leads to a very important relationship between the vertical variation of geostrophic wind and the horizontal temperature gradient, or the so-called *thermal wind relation*.

Consider a region of the atmosphere with pressure and temperature distributions as shown in Figure 14. For simplicity, we shall assume here that the temperature gradient

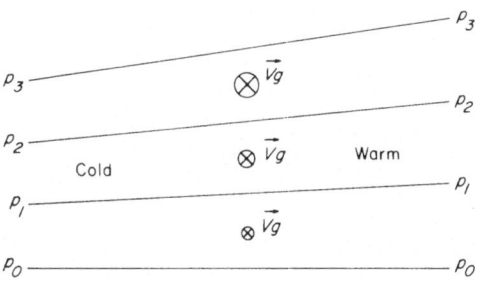

Fig. VII-14. *Variation of geostrophic wind with height.* The thickness of an atmospheric layer between two isobaric surfaces is proportional to the mean temperature of the layer. If there is a horizontal temperature gradient, the isobaric surfaces are sloped downward from the warm region to the cold region. This slope increases with height because successive layers of air are always thicker in the warm region. This implies a vertical increase in the horizontal pressure gradient, and thus a vertical increase in the geostrophic wind. In this simple illustration, we have assumed that the surface pressure is uniform and that the horizontal pressure and temperature gradients are in the same direction.

and the pressure gradient are in the same direction. Let $z_0, z_1, z_2, \ldots$ be the altitudes of the isobaric surfaces $p_0, p_1, p_2, \ldots$ respectively. The thickness of the layer between two isobaric surfaces p_{i+1} and p_i is $(z_{i+1} - z_i)$. Its value should be larger in the warm region than in the cold region for all $i = 0, 1, 2, \ldots$. The slope of the isobaric surface should therefore increase with height. As a consequence, the horizontal pressure gradient also increases with height. This conclusion could be demonstrated rigorously, but a simple examination of Figure 14 makes it already apparent.

Hence the geostrophic wind should also increase with height. The increase of geostrophic wind with height is in the direction perpendicular to the temperature gradient, with the cold region to the left-hand side and warm region to the right-hand side of the vector of wind increment.

In general, the horizontal temperature gradient needs not be parallel to the horizontal pressure gradient. However, the vertical change in the geostrophic wind, called the *thermal wind*, is always perpendicular to the horizontal temperature gradient with the colder region to the left-hand side of the thermal wind vector. If we designate the thermal wind vector between the pressure levels p_0 and p_1 ($p_1 < p_0$) by $\mathbf{v}_T$, we have, by definition,

$$\mathbf{v}_T = \mathbf{v}_g(p_1) - \mathbf{v}_g(p_0) \tag{46}$$

It can be shown by straightforward differentiation of the geostrophic wind relation (41) that

$$\mathbf{v}_T = -\frac{R}{f} \int_{p_0}^{p_1} (\hat{z} \times \nabla T)\,\mathrm{d}\ln p = -\frac{R}{f}(\hat{z} \times \nabla \bar{T}) \ln \frac{p_1}{p_0} \tag{47}$$

where R is the gas constant and $\bar{T}$ is the mean temperature of the layer $p_1 < p < p_0$. The relationship between the thermal wind and the horizontal mean temperature gradient is demonstrated in Figure 15.

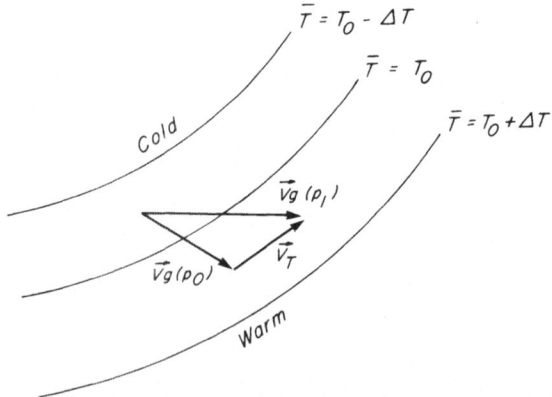

Fig. VII-15. *Thermal wind*. The thermal wind $\mathbf{v}_T$ between two isobaric surfaces p_0 and p_1 (with $p_0 > p_1$) is in a direction perpendicular to the horizontal gradient of the mean air temperature of the layer, with the cold region to the left of the thermal wind vector. The geostrophic wind on the p_1 surface $\mathbf{v}_g(p_1)$ is related to the geostrophic wind on the p_0 surface $\mathbf{v}_g(p_0)$ through $\mathbf{v}_g(p_1) = \mathbf{v}_g(p_0) + \mathbf{v}_T$. In the case shown, the geostrophic wind has a component going from the cold region to the warm region. The geostrophic wind rotates with altitude in the counter-clockwise direction. If the geostrophic wind blows from the warm region to the cold region, it will rotate with altitude in the clockwise direction.

8. The Thermal Circulation

An atmospheric circulation may be induced if there is an uneven distribution of heating and cooling rates in the atmosphere. This is called a *thermal circulation*. In a thermal circulation, the air motion is initiated and maintained by the pressure gradient force caused by the uneven temperature distribution. Many atmospheric circulations can be understood in these terms.

Because of the rotation of the Earth, pure thermal circulation is rarely observed in systems with large spatial and temporal scales. For simplicity, however, we shall first discuss in this section the properties of a thermal circulation without the influence of the Coriolis force.

Consider a region of the atmosphere that initially has horizontally homogeneous distributions of temperature and pressure as demonstrated in Figure 16(a). Let us suppose that, due to certain mechanisms, heat is added to the air at one end of the region and extracted at the other end. Because of these heating and cooling processes, the internal energy of the air is increased in one region and decreased in the other. If the heating and cooling processes are slow and are realized over a very large area, the state of hydrostatic equilibrium of the atmosphere will not be disturbed.

As has been discussed in §6.1, the total internal energy of an air column in hydrostatic equilibrium is proportional to its total potential energy. When an air column is heated to increase its internal energy, the air column must expand vertically so that its potential energy will be raised proportionally. Similarly, an air column should contract vertically in the region where it is being cooled. As a result, a pressure gradient as indicated in Figure 16(b) should appear between the warm and cold regions.

As soon as a horizontal pressure gradient appears between the warm and cold regions, the pressure gradient force will force the air to flow in the direction from the warm area to the cold area. This air flow will then tend to decrease the pressure in the warm region and to increase the pressure in the cold region because the pressure at a point in a hydrostatic atmosphere is equal to the weight of air above the point.

The rate of decrease of pressure at a given height in the warm region is equal to the integrated rate of air mass outflow above the height. Therefore the rate of pressure decrease due to the outflow of air in the warm region should be larger in the lower levels than in the upper levels. For the same reason, the rate of increase in pressure in the cold region should be larger in the lower levels than in the upper levels.

Eventually a steady state as shown in Figure 16(c) will be reached. At a given height, the pressure will be lower in the warm region and higher in the cold region at the lower levels. At the upper levels, the pressure will be higher in the warm region than in the cold region. The direction of air flow in the lower levels should also be reversed as shown. This is a basic pattern of thermal circulation.

An example of thermal circulation is the land or sea breeze. During the daytime, because of its small heat capacity, the land surface temperature will usually rise higher than that of the adjacent sea surface. This higher land surface temperature will lead to the air over land being warmed more than that over the sea. A thermal circulation will be initiated with the air blowing from the sea to the land near the surface; this surface wind carries the cool and moist air from the ocean over the land and is called the sea breeze. It is often felt as a refreshing relief during insolation hours in hot climates, along coastal areas.

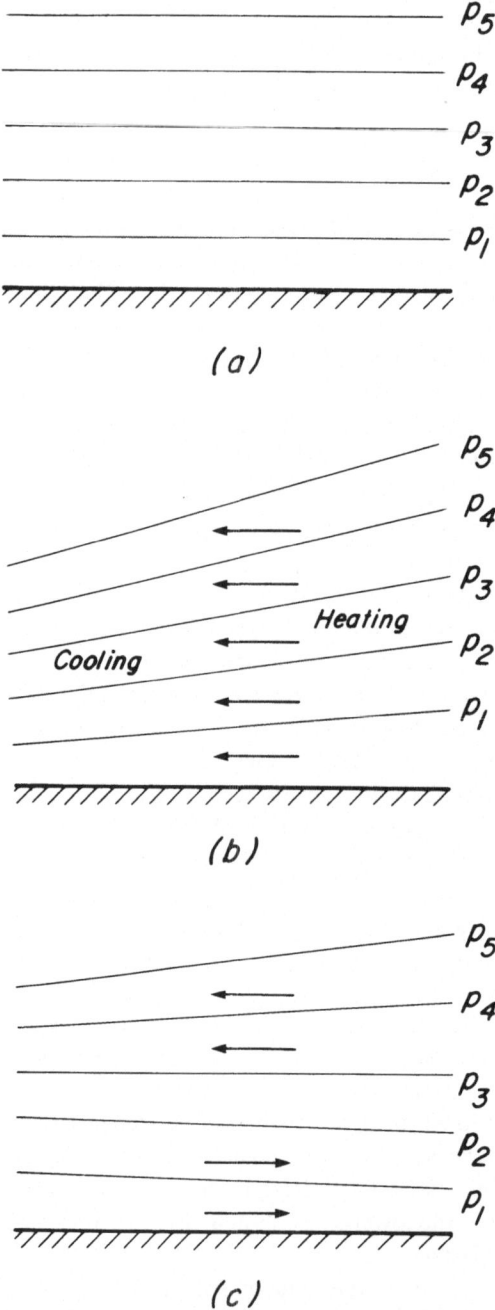

Fig. VII-16. *Thermal circulation.* When heating and cooling processes are applied to different parts of a homogeneous region of the atmosphere shown in (*a*), an air column will expand in the heated region, and will contract in the cooled region. A pressure gradient will appear to induce an air flow from the warm region to the cold region as shown in (*b*). The air flow will then readjust the hydrostatic pressure distribution to reach a possible steady state configuration as shown in (*c*). In this state, the air flows from the cold region to the warm region in the lower levels, and in the opposite direction in the upper levels.

At night, the contrast between the land and sea surface temperature is reversed. A thermal circulation cell in the reverse direction then appears and the surface wind is now a land breeze.

Land and sea breezes are most likely to occur when the sky is clear and the wind is otherwise light, so that maximum land and sea temperature contrast can develop. Typical sea breeze and land breeze circulations are illustrated schematically in Figure 17.

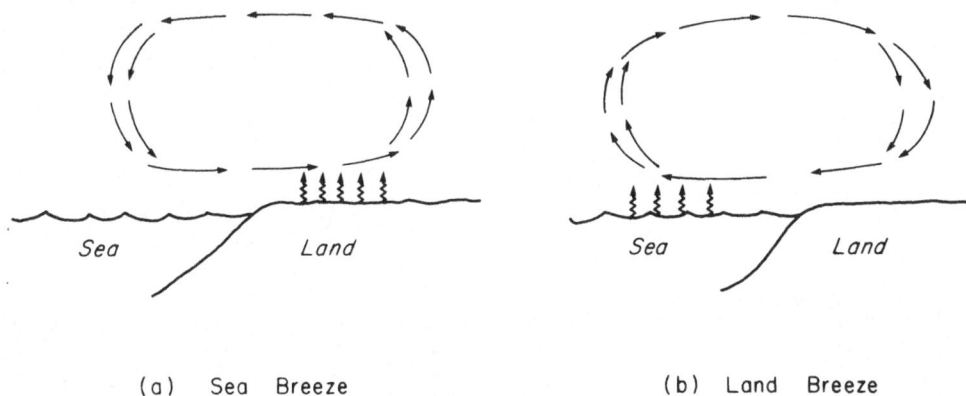

(a) Sea Breeze   (b) Land Breeze

Fig. VII-17. *Sea and land breeze*. Thermal circulations are often induced by the land and sea surface temperature contrast in coastal regions. During daytime, land surface is often warmer than sea surface. A thermal circulation can be induced with the surface wind blowing from sea to land. This is the sea breeze (*a*). During night-time, the temperature contrast is reversed. A thermal circulation in the opposite direction can be induced, with the surface wind blowing from land to sea. This is the land breeze (*b*).

9. The General Circulation of the Atmosphere

The pattern of atmospheric motion at any given instant always shows a great deal of complexity. However, the motion of air over the Earth has certain characteristic gross features, but these are often concealed by the superposition of many smaller scale elements of air motion. In this section we shall discuss a few of the important gross features of the atmospheric global circulation.

In order to make the discussions possible, we must first separate the global-scale features from the small-scale features. It should be understood, however, that the atmosphere is a highly nonlinear system; there are always interactions between motion systems of different scales. As a consequence, one cannot separate physically the large-scale motion systems from smaller-scale systems, and any separation introduced for the sake of a better understanding carries with it a certain degree of artificiality or arbitrariness.

One method to remove the small features is to perform longitudinal or time averages. Small-scale motions usually have only very limited longitudinal extent. They also exist usually only for a relatively short time period. The complexity in the pattern of air motion caused by the presence of these small features may be 'removed' by averaging the motion fields over a longitudinal circle and/or over an extended period of time, say a month. These longitudinally-averaged and time-averaged fields are the main subject to be discussed in this section.

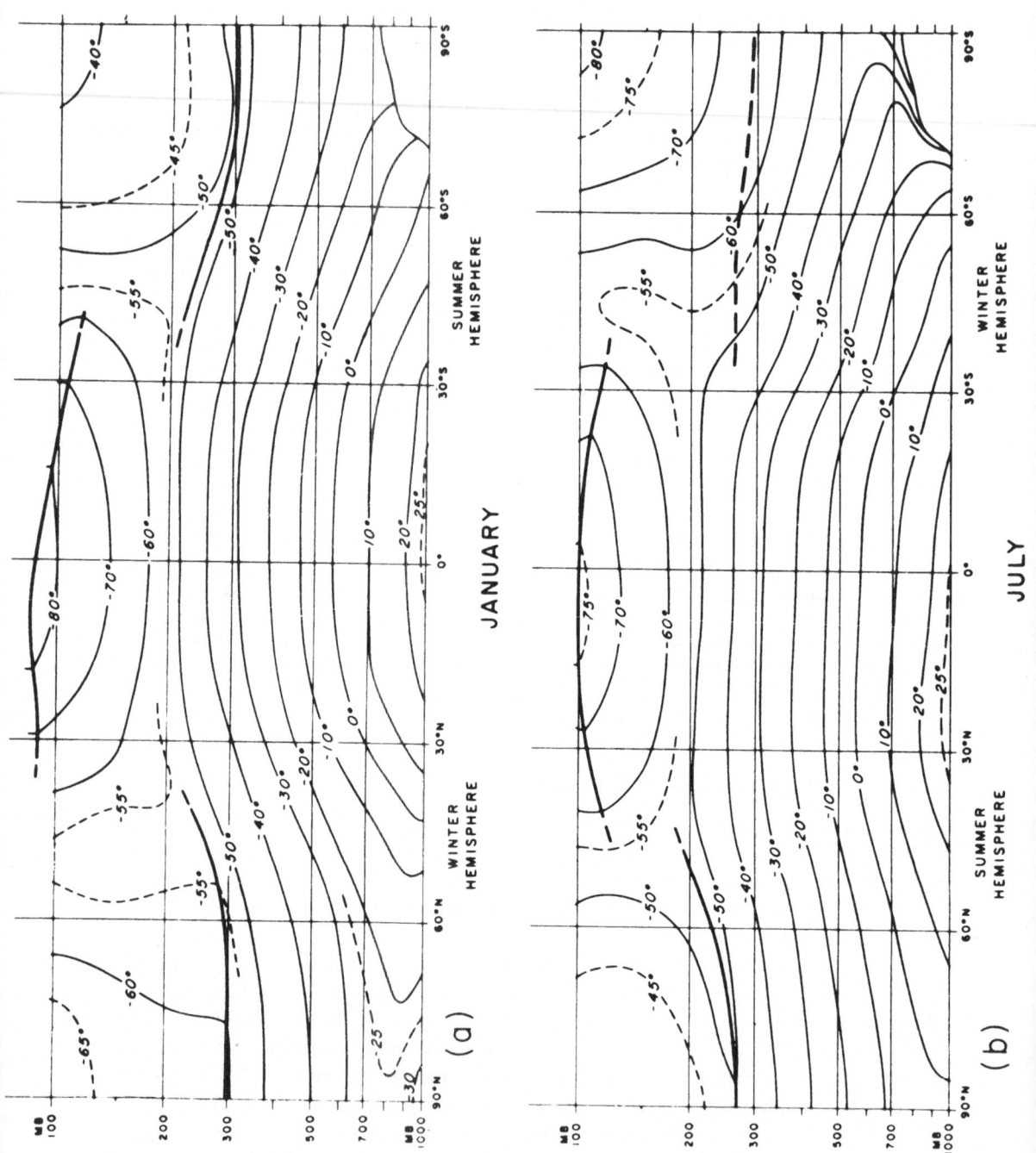

Fig. VII-18. *Zonally-averaged temperature distribution,* for (*a*) January and (*b*) July. The temperatures are in degrees Celsius. Heavy lines mark the position of the tropopause. The north–south temperature gradient is strongest in mid-latitude regions. It is stronger in the Winter Hemisphere than in the Summer Hemisphere.

174

By doing this, however, we are not implying that the mean circulation field may exist independently of the smaller-scale systems. Many smaller systems are in fact an integral part of, and inseparable from, the mean global circulation. This will also be discussed in this section.

9.1. *The longitudinally-averaged temperature distributions*

The distribution of mean temperature averaged around latitude circles for the months of January and July are shown in Figure 18. The heavy lines mark the position of the tropopause. Overall, there is a poleward decrease in mean temperature at all levels, except above about the 200 mb pressure level in the tropics ($30°N$–$30°S$).

Qualitatively, the pole-to-equator temperature gradient can be understood simply by considering the energy balance of the atmosphere. The Sun is the only energy source for the Earth's atmosphere. The solar radiational energy absorbed by the Earth–atmosphere system has a strong latitudinal dependence. There is a net radiational energy surplus in the lower latitudes around the equator, and a net radiational energy deficit in the middle and higher latitudes (cf. Ch.III, §10). Consequently, the mean air temperature in the tropical region should be higher than that in the middle and high latitudes. This north–south temperature gradient is the main driving force of the global circulation of the atmosphere.

A quantitative explanation of the observed temperature distribution, however, is quite a difficult matter. Many detailed features cannot be easily understood even qualitatively. The horizontal gradient of the mean temperature is very weak in the tropical region between $30°N$ and $30°S$. Most of the equator-to-pole temperature decrease occurs in the mid-latitude regions ($30°N$–$60°N$, and $30°S$–$60°S$). The temperature gradient is much stronger in winter than in summer. None of these features can be explained by the radiational energy balance alone. The transport of heat energy produced by the motion of air itself plays a very important role here.

9.2 *The Hadley circulation*

The first model to describe the global air circulation pattern was proposed by G. Hadley in 1735. The *Hadley circulation* is essentially a direct thermal circulation as described in §8.

Based on the observation that air in the lower latitudes is warmer than the air in higher latitudes, tropical air should rise vertically and move northward in the upper troposphere, while the cold polar air should sink and move southward in the lower troposphere. The northward moving warm tropical air will lose much of its heat energy through radiative cooling before it reaches the polar region to replace the descending and southward-moving cold air. The cold air from the pole will absorb heat from the ground (itself radiationally heated) in the lower latitudes and then ascend in the equatorial zone. The essential features of the Hadley circulation is illustrated in Figure 19.

A thermal circulation of this type is obviously capable of transporting thermal energy poleward to balance out at least part (and, it is hoped, all) of the radiational energy surplus in the equatorial region and the radiational energy deficit in the polar region. The model, however, has some serious dynamical defects.

175

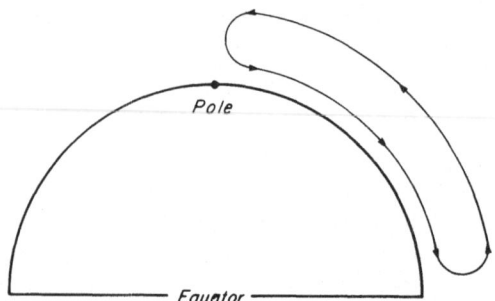

Fig. VII-19. *The Hadley cell.* The first model proposed in 1735 by G. Hadley to describe the global mean circulation. Air rises slowly in the warm tropical region and moves northward, losing thermal energy by radiation, then descends over the cold polar region and returns to the lower latitudes in the lower atmosphere near the Earth's surface.

The typical pressure distribution associated with a thermally-driven circulation has been discussed in §8. According to that picture, there should be an equator-to-pole pressure gradient (and therefore a pole-to-equator pressure gradient force) in the lower troposphere and a pole-to-equator pressure gradient in the upper troposphere.

The rotation of the Earth tends to deflect the wind toward the direction parallel to the isobars so that a geostrophic balance is approximately maintained. This implies that the wind in the troposphere should have a strong easterly component in the lower layer and a strong westerly component in the upper layer. Due to the frictional force between the Earth's surface and the lower troposphere, the easterly wind in the lower atmosphere means a constant transfer of westerly momentum from the Earth to the atmosphere, or a transfer of easterly momentum from the atmosphere to the Earth. Such a constant exchange of angular momentum between the Earth and the atmosphere is not allowed in a steady state. If the mean wind velocity in the lower atmosphere has an easterly component in one region, it must have a westerly component in some other region, so that the net angular momentum exchange between the Earth and the atmosphere is zero.

9.3. *The observed mean global circulation*
There are substantial differences between the Hadley model and the observed global circulation patterns.

The observed mean surface pressure and wind distributions over the Earth are represented schematically in Figure 20. Over the equator there is a low pressure belt (*equatorial low pressure belt*); at about 30°N and 30°S, there are the so-called *subtropical high pressure belts*. Between the equatorial low pressure belt and the subtropical high pressure belt the wind is north-easterly in the Northern Hemisphere and is south-easterly in the Southern Hemisphere. They are called the *North-East* and *South-East trade winds*, respectively. This part of the circulation was very important to the early navigation based on sailing ships; hence the name of 'trade' winds.

Between the subtropical high pressure belts and the two low pressure belts which can usually be found at around 60°N and 60°S, the winds are prevailing westerlies. The south and the north poles are usually two high pressure regions. The winds in the polar regions are usually easterlies.

176

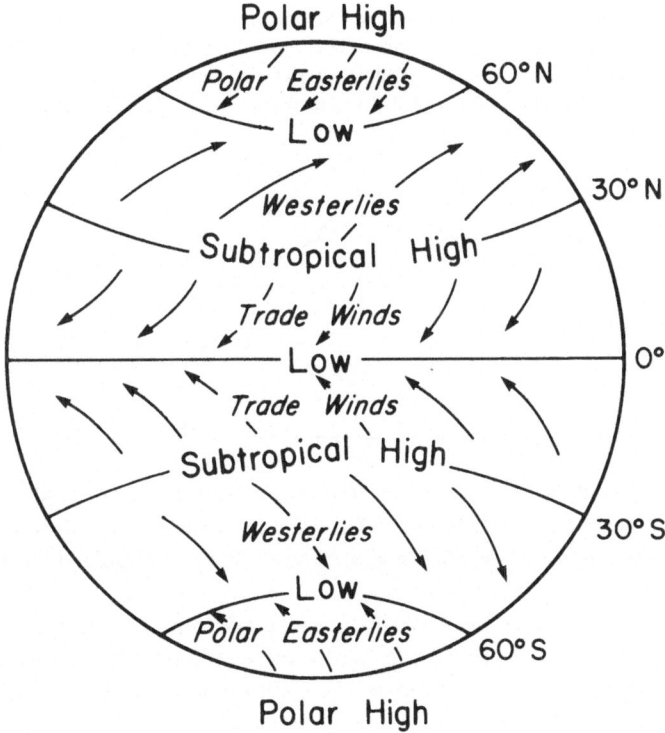

Fig. VII-20. *Surface wind and pressure distributions*. Schematic illustration of the observed mean surface wind and pressure distributions.

The pressure and wind distributions shown in Figure 20 are meant only as a schematic representation. The locations of the boundaries between the various regions have large seasonal variations. On any particular day, the wind and pressure distributions also have large east–west (zonal) variations.

The equatorial low pressure belt and the polar high pressure centres are what we should expect if the Hadley model were an adequate description of the global circulation. The subtropical high pressure belts and the lower pressure belts at 60°N and 60°S are, however, unexpected. A thermally-direct circulation cannot possibly explain the observed surface pressure distribution. The observed surface wind distribution is the result of the balance of the pressure gradient force, the Coriolis force, and the surface frictional force. The alternating belts of easterlies and westerlies ensure that there is, on the average, no net exchange of angular momentum between the Earth and the atmosphere.

In Figure 21 we show schematically the *meridional circulation* pattern of the atmosphere in the Northern Hemisphere. The direct thermal circulation with air ascending in the equatorial region, as suggested by Hadley, does exist and is called the Hadley cell. The poleward extension of this cell reaches, however, only about 30°N. Between 30°N and 60°N, where the north–south temperature gradient is the strongest, the mean meridional circulation is opposite in direction to that required by the thermal circulation. This circulation is called the *Ferrel cell*. In the Ferrel cell, air rises in the colder region around 60°N and descends in the warmer region around 30°N.

Fig. VII-21. *Mean meridional circulation.* Schematic illustration of the vertical cross section of mean meridional circulation of the Northern Hemisphere. There are three circulation cells. The thermally-direct Hadley cell extends only to about 30°N; the Ferrel cell is thermally indirect and occupies the region between 30°N and 60°N; the polar cell is a rather weak circulation. The arrows on the hemispherical semi-circle indicate surface wind directions.

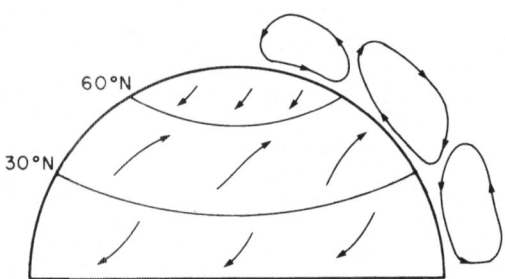

In addition to the Hadley and the Ferrel cells, a third cell exists over the polar region. This polar cell is a very weak circulation. Air ascends in the cell in the warmer region around 60°N and descends in the cold region around the north pole. Like the Hadley cell, this polar cell is a thermally-direct circulation.

Figure 22 shows the cross sections of the mean meridional circulations constructed from real observational data for the winter and summer seasons in the Northern Hemisphere. There are large seasonal variations in the position and the strength of the circulation cells. The Hadley cell is much stronger in the winter than in the summer season. During the summer season, the Hadley cell is moved northward into the region between 15°N and 45°N. The ascending branch of the Hadley cell of the Southern Hemisphere extends as far north as 15°N.

There is also a seasonal variation in the position of the Ferrel cell. During the summer season, the Ferrel cell is located in the region between 45°N and 65°N; during the winter season, it occupies the region between 35°N and 75°N. The polar cell is not represented in the figures because observational data in the polar region are very scarce and the polar cell is a very weak circulation.

The distribution of the longitudinally-averaged *zonal wind* (east–west wind) during the winter and summer seasons for both hemispheres is illustrated in Figure 23. The surface wind is easterly between 35°N and 35°S. This is the region of the trade winds, associated with the returning branch of the Hadley cells. The surface wind is westerly in the middle latitude regions occupied by the Ferrel cells, and it becomes easterly again in the polar regions to the north of 65–70°N and to the south of 65–70°S.

The upper level zonal winds are mostly westerlies in the troposphere, except in the region between 15°N and 15°S, around the equator. In the middle latitudes, the westerly wind increases steadily with height until a maximum is reached at an altitude just below the tropopause. This feature is associated with the strong north–south temperature gradients in these regions. As discussed in §7.2, the change in the geostrophic wind – called the thermal wind – is caused by the horizontal gradient in temperature. The thermal wind vector should be perpendicular to the horizontal temperature gradient, with the cold region to the left of the thermal wind. Since the north–south temperature gradient is strongest in the middle latitudes, the increase in the westerly wind component is most apparent in this region.

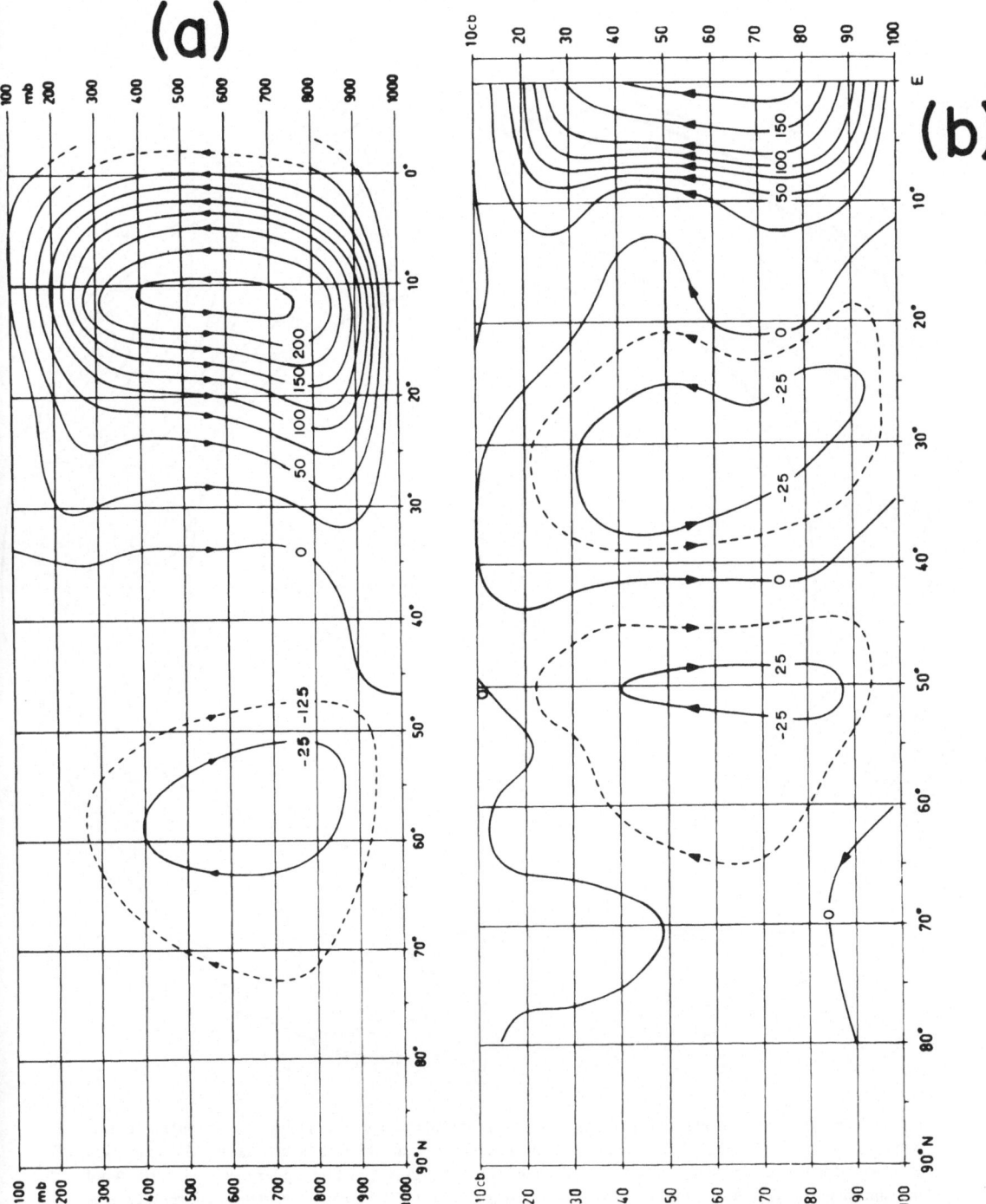

Fig. VII-22. *Observed mean meridional circulation.* The zonally-averaged meridional circulations in the Northern Hemisphere based on observations are shown for (*a*) the winter season and (*b*) the summer season. The figures associated with each streamline are in units of mass flux. Each streamline channel transports mass at a rate of 25×10^6 ton/sec. The Hadley cell is much stronger during the winter season than during the summer season. During the summer season, the ascending branch of the Hadley cell of the Southern Hemisphere moves into the Northern Hemisphere.

179

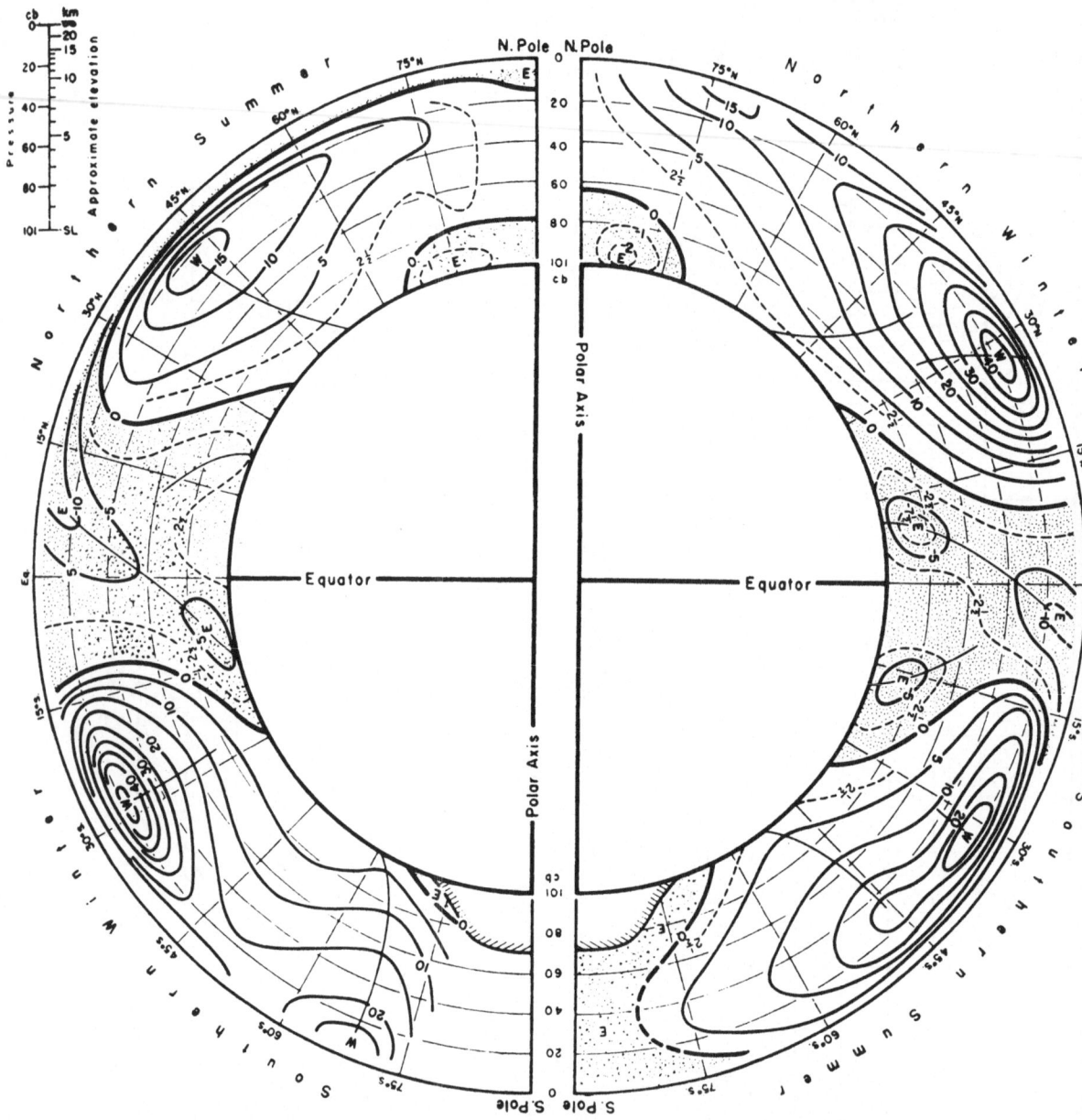

Fig. VII-23. *Mean zonal wind distribution*. Longitudinally-averaged zonal wind distribution in the winter and summer seasons. Positive values denote westerly wind. The surface zonal wind distribution is essentially the same as demonstrated schematically in Figure 20. The upper level westerly wind maximum is stronger in winter than in summer. It is caused by the stronger north–south temperature gradient in the winter season. cb = centibar = 10 mb.

From the observationally-based descriptions just presented, an interesting picture of the mean atmospheric circulations has emerged. Many isolated and seemingly unrelated phenomena observed in the atmosphere can be now fitted into the integrated picture. At the same time, the picture also presents a few very puzzling problems to be resolved.

The very steady N.E. trade winds and S.E. trade winds, so important to early-day sailors, can now be understood as the surface parts of the returning branches of the Hadley cells. The Earth's surface in the tropical region is mostly covered by the oceans; as the air in the trade wind regions flows toward the equator, it collects sensible heat and water vapour from the ocean surface. A moist layer of air is established in the lower troposphere. The vertical stratification in this moist layer usually is conditionally unstable (cf. Ch.IV, §9). Moist convection in the form of cumulus clouds can usually be observed.

As the air in trade winds flows towards the equator, more and more moisture is accumulated in the moist layer; the layer deepens, and the cumuli grow taller as well. When the air reaches the ascending branch of the Hadley cell, it usually has a very high moisture content; this, when coupled with the steady, ascending motion of air, produces many deep cumulonimbus clouds. A narrow east–west cloud band can usually be identified from pictures taken from satellites, especially over the Atlantic and Pacific Oceans. This is the region known as the *Inter-Tropical Convergence Zone* (ITCZ).

Because of the asymmetry of the ocean and land distributions between the Northern and Southern Hemispheres, the ITCZ is usually located in the region between the equator and 10°N. One example is shown in Figure 24. It is a picture taken in the visible region of the radiation spectrum by the European Space Agency's Meteosat Satellite. The cloud band is clearly visible over the Atlantic Ocean between the west coast of Africa and the east coast of South America.

Since the Hadley cells are thermally-direct circulations, they serve to transport heat energy poleward to balance out, in part at least, the radiational energy deficit at the higher latitudes (cf. Ch.III, §10). This transport mechanism reaches, however, only 30°N and 30°S; in the middle latitude regions, the heat energy transport must be taken over by some other mechanisms.

The Ferrel circulations over the mid-latitudes clearly do not serve this purpose. The Ferrel cells are circulating in a direction opposite to that required by the thermal circulation. The poleward energy transport in these regions is carried out by the *eddy motions* which are constant features of the atmospheric global circulation.

9.4. *The eddy motions*

The term '*eddy*' is usually used in meteorology to describe the departure of air motion from the mean circulation described previously in this section. Eddy motions are very important components of the atmospheric global circulation. There are many different kinds of eddies, with scales ranging from the planetary scale down to the small scale.

It is well known that, even in the very steady trade wind regions, storms do occur from time to time. These storms are usually associated with intense cumulus convective activities, which produce heavy precipitation. Occasionally, some of these storms may intensify to develop into hurricanes or typhoons. These eddies in the tropical and subtropical regions play a very important role in both tropical and global energy balances.

There are also a variety of eddies in the middle latitude regions. Some of these are rather stationary on a seasonal basis; others are more transient in nature. Eddies of the

Fig. VII-24. *Intertropical convergence zone (ITCZ)*. In this full view of the Earth, the east–west cloud band over the Atlantic Ocean between the west coast of Africa and the east coast of South America is the ITCZ. (Photo by Meteosat 1/ESA.)

planetary scale and of the synoptic scale are very important in the poleward thermal energy transport in the mid-latitude regions. To demonstrate how eddy motions may transport thermal energy, let us consider the following simple example.

Suppose that an eddy motion in the Northern Hemisphere has stream lines on a horizontal plane as shown in Figure 25. The parts where the stream lines approach closest to

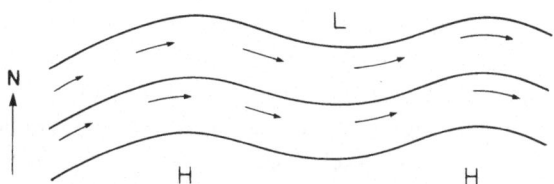

Fig. VII-25. *Eddy motion*. A wave-like eddy motion with ridges (the parts where streamlines approach closest to the pole) and troughs (the parts where streamlines approach closest to the equator). A northward thermal energy transport can be accomplished without a net mass transport if the northward-moving air parcels in the eddy are warmer than the southward-moving parcels.

the pole are called ridges, and those where they are nearest to the equator are called troughs. To the left of the ridges, air moves eastward and northward; to the right of the ridge, air moves eastward and southward. If, for some reason, the air at the left side of the ridges is warmer than the air at the right side, we have a situation in which warm air is moving northward and cold air is moving southward. The net effect is a northward transport of thermal energy without any net northward transport of mass.

The difference in air temperature between the right and left sides of the ridges can be caused by a variety of things. One mechanism, for example, is that air near the ridges may lose heat energy by radiation while air near the troughs picks up thermal energy from the warmer earth surface in the southern regions.

The mid-latitude eddy motions can be established by several different mechanisms. One of them is the seasonally-contrasting thermal influences of land and ocean surfaces. During the summer season, the land surface temperature is – due to its small heat capacity – higher than the temperature of the ocean surface. Consequently, a thermally-driven circulation can be set up between the land area and the ocean area. A surface low pressure centre then appears over land and a surface high pressure center over the ocean. In the winter season, the land and ocean surface temperature contrast is reversed. Consequently, low pressure cells tend to be centred over the oceans and high pressure cells over large land masses. The eddies produced by this mechanism usually have very large horizontal dimensions comparable to the dimensions of large oceans and continents. They are usually of the planetary scale and also tend to be rather stationary during the winter and summer seasons.

In addition to the seasonal semi-permanent cells, there are also migrating low and high pressure cells, of synoptic scale, usually transient in nature, that also contribute importantly to the exchange of cold and warm air masses between the subtropical region and polar regions. The formation process of these migrating cells is basically dynamical in nature. It is usually associated with the development of zones of sharp thermal contrast (called *fronts*). Because of the importance of this type of disturbance to the daily weather changes in the middle latitudes, we shall discuss its properties in a separate section.

10. Air Masses and Fronts

Air mass can be defined as a large body of air whose properties, such as temperature and humidity, are more or less homogeneous on a horizontal plane. This concept is very useful to underline the most important features of practically-unlimited weather situations, and plays a very important role in daily weather forecasting.

The properties of air depend on the radiation processes and on the exchange of heat and moisture with the underlying Earth surface. These processes are usually very slow. It takes several days for them to substantially modify the properties of the air.

The radiation process is different at different latitudes. The energy exchange with the Earth depends on many surface features. The way in which air is modified by these processes should therefore depend on its geographic location. If a mass of air has stayed at one particular region for a sufficiently long period of time, it tends to take on the characteristics defined by the location. For example, an air mass that has been staying for a certain time over the polar region tends to be cooler than a mass of air that has been located in the tropical area. The air mass from over the ocean is usually more moist than one from a large continent.

When an air mass is moved quickly from its 'source region' into a different location by some dynamical process, a sharp contrast in air properties can usually be found between the air mass and its surrounding air. Since the properties of each air mass as a whole change only very slowly, most of the transitions of properties between the two air masses occur in a narrow region at the common boundary. This boundary, which is usually characterized by strong gradients of air properties, is called a *front*. Often, one can make a rough weather forecast by predicting the motions of the air masses.

There are many different ways to classify the various types of air masses. Usually, an air mass falls into one of the three principal classes: *tropical air mass*, *mid-latitude air mass*, and *polar air mass*, depending on its source region. Air masses of each class can usually be further classified into two subclasses: *continental air mass* and *maritime air mass*, depending on whether the source region is over a continent or over an ocean.

The source regions of tropical air masses are the trade-wind zones of the Northern and Southern Hemispheres. The Earth surface in this region is occupied mostly by the oceans. Air masses over this region collect large amounts of heat and moisture from the ocean surface; thus tropical air masses are often characterized by their warm temperatures and high moisture content.

The principal source regions of polar air masses are the arctic and subarctic regions in the Northern Hemisphere and the antarctic and subantarctic regions of the Southern Hemisphere. Because of the large radiational energy loss in these areas, the polar air masses are characterized by their very low temperatures.

The mid-latitude air masses originate in the mid-latitude westerly regions. The surface features in the westerly region in the Northern Hemisphere have large zonal variations. Because of the relatively large extensions of land, mid-latitude air masses originating over continents often have quite different characteristics from those originating over the oceans. Their properties tend to be less homogeneous than those of the tropical and polar air masses.

The boundary between the polar and the mid-latitude air mass is called the *polar front*. The polar front is characterized by very strong north–south temperature gradients. Figure 26 gives the meridional cross section showing mean conditions around the Northern Hemisphere on January 1, 1956. The position of the polar front at the surface at this time is about 30°N, and it slopes northward with height. At the top of the polar front, there is a region of extremely strong westerly wind; it is called the *polar jet stream*. The

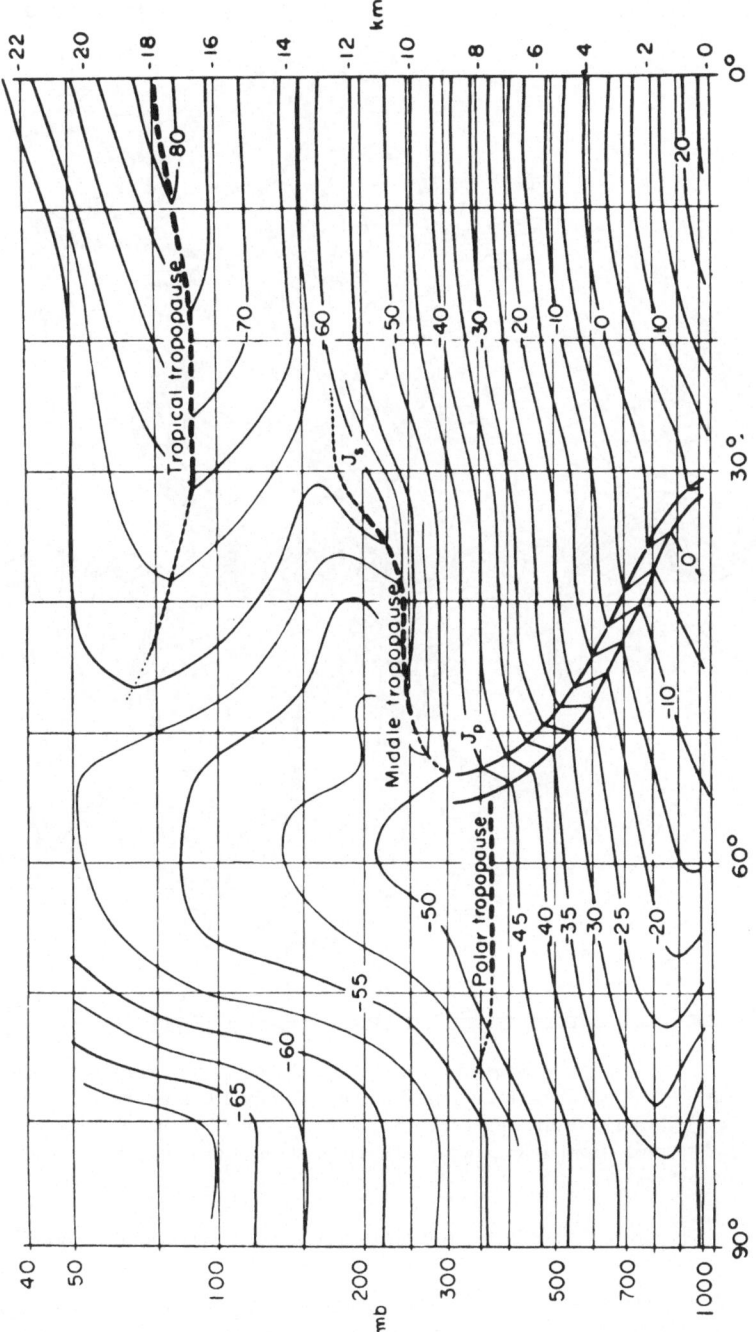

Fig. VII-26. *Polar front and subtropical front.* Meridional cross section showing zonal mean temperature distributions around the Northern Hemisphere on 1 January 1956. The averaging is done relative to the principal zones of north–south temperature gradient. Isotherms are in degrees Celsius. J_p and J_s mark the positions of the polar and subtropical jet streams. The subtropical front is very diffused in comparison with the polar front.

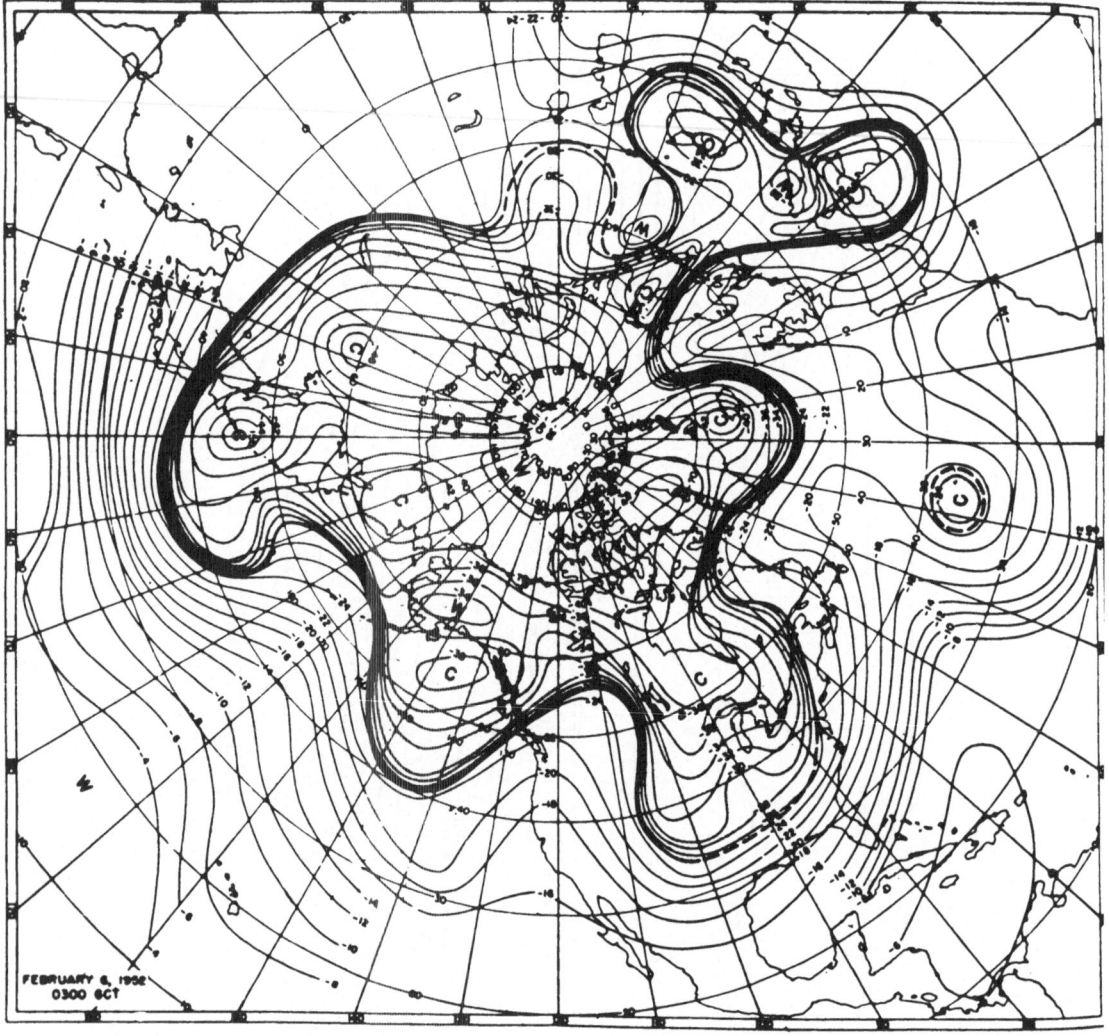

Fig. VII-27. *Zonal variation of the polar front.* The figure shows the temperature distribution on the 500 mb surface at 03 GCT (Greenwich Central Time), 6 February 1952. Isotherms are drawn at 2°C intervals. The heavy cluster of isotherms marks the southern limit of the polar air, i.e. the position of the polar front. The polar front is somewhat diffused at several locations, due to the presence of smaller-scale disturbances.

polar jet stream is caused mainly by the large north–south temperature gradient across the front.

At a given time, the structure of the atmosphere always has large zonal variations. The position of the polar fronts often vary with longitude. Figure 27 shows the temperature distribution on the 500 mb surface at 03 GCT on February 6, 1952. The heavy cluster of isotherms marks the position of the polar front on this surface. The position of the polar front shows a wave-like structure with a maximum southward displacement to 30°N and maximum northward displacement to 70°N. The polar front is also broken at a few places where it is somewhat diffused; this feature is caused by the presence of eddy motions.

186

The boundary region between the middle-latitude air mass and the tropical air mass is called the *subtropical front*. The subtropical front usually is not as sharply defined as the polar front. As can be seen from Figure 26, the temperature gradient across the subtropical front is somewhat diffused. The location of the front is often more easily located by the *subtropical jet stream* at the top of the frontal system. This strong westerly wind jet stream is, of course, also caused by the strong north–south temperature gradient associated with the front.

The wind speed in the jet stream can be as high as 100 m/s. This is a non-negligible factor in air navigation, even for modern jet aeroplanes; it can be used to shorten flying time when travelling eastwards.

11. The Mid-latitude Cyclones

The state of the atmosphere on any given day at a location in mid-latitude regions usually deviates substantially from the mean Ferrel cell circulation. We have already discussed in §9 the reasons why eddy motions in these regions are essential components of the atmospheric global circulation. In this section, we shall discuss the dynamical properties of one type of eddy, namely the synoptic scale low pressure systems (or cyclones), which are mostly responsible for the daily weather variations in middle latitudes.

The synoptic scale mid-latitude cyclones usually develop along the polar front where the cold air mass from the polar region and the warm air mass of a tropical nature meet. The formation of these disturbances is caused by an instability process known as the *baroclinic instability*.

To explain this baroclinic instability process in very simple terms, let us consider a region of the atmosphere where a strong north–south temperature gradient exists. This region is schematically represented in Figure 28; here, the sloped surfaces are the

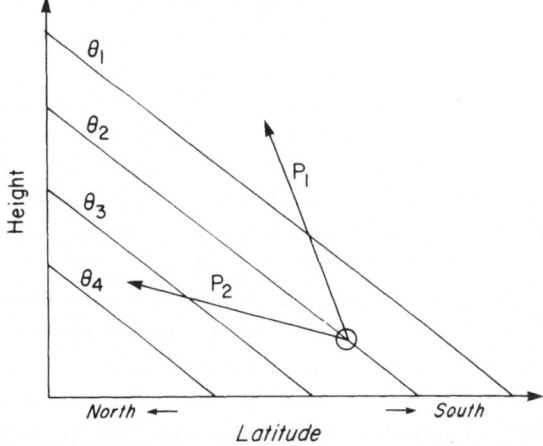

Fig. VII-28. *Baroclinic instability*. A region of strong north–south temperature gradient is shown. θ is the potential temperature, with $\theta_1 > \theta_2 > \theta_3 > \theta_4$. An air parcel in such a region becomes unstable if it is pushed northward along path P_2. Along such a path, it is surrounded by colder air. The parcel will continue to rise and to move northward. On the other hand, if the parcel moves along path P_1, it enters a region of warmer air. The parcel tends to sink back to its original height. It becomes a stable situation.

constant potential temperature surfaces (cf. Ch.IV, §5), with potential temperatures $\theta_1 > \theta_2 > \theta_3 > \theta_4$. If, by some process, a parcel of warm air from the south is moved adiabatically to the north, it enters into a region where it is surrounded by colder air. Since warmer air has a lower density, the parcel of warm air tends to rise as it moves northward. Depending on its rate of ascent, the parcel will follow either one of the two types of paths P_1 and P_2 shown in the figure. If the parcel moves long path P_1, it will enter into a region where it is surrounded by warmer air; the parcel of air should then sink back towards the surface of its original potential temperature. On the other hand, if the parcel moves along path P_2, it will always be surrounded by colder air and, hence, will continue to rise. The surrounding cold air should then descend to take the place of the ascending warm air. Since warm air has a lower density, the centre of gravity of the atmosphere is lowered as the warm air rises and the cold air sinks. The loss of potential energy during this process is converted into the kinetic energy of the air motion and the disturbance intensifies.

The rising air is usually of a tropical origin, and often has a high moisture content. As the air rises, its temperature decreases; the water vapour contained in the air quickly becomes saturated and begins to condense, clouds of various forms begin to form and produce precipitation. The disturbance is, therefore, invariably associated with 'bad weather'.

The above discussion, although it gives a simple illustration of the dynamical origin of mid-latitude cyclones, actually constitutes a drastic oversimplification of the complex dynamical processes leading to the initiation and intensification of low pressure systems and the associated cyclonic air motions. Instead of presenting a detailed dynamical explanation of these processes, which would be beyond the scope of this book, we shall give only a qualitative description of the development of such systems.

First, we shall explain a few terms and symbols that we intend to use in the following discussion.

As we have discussed before, a front is the boundary region between two air masses of different properties (temperature, humidity, etc.). Here we shall consider a warm air mass and a cold air mass. If air from the warm air mass moves toward and glides over the cold air mass, the boundary region is called a *warm front*. On the other hand, if the cold air moves into the warm air, the boundary is called a *cold front*. If the front does not move, it is called a *stationary front*. The conventional symbols used to mark the positions of a cold, warm, or stationary front are illustrated in Figure 29(a), (b) and (c) respectively. The symbol used for a fourth type of front, known as the *occluded front*, is shown in Figure 29(d); its meaning will be explained later in this section.

The mid-latitude cyclones usually go through well defined life cycles. The disturbances usually form at the quasi-stationary polar front. We shall assume here that the wind to the south of the front is westerly and the wind to the north is easterly, as shown in Figure 30(a). When a perturbation caused by the northward motion of the warm air appears, the front becomes curved as shown in Figure 30(b). The pressure drops at the northern tip of the frontal system. The colder air moves out southeastward at the surface into the region which is occupied by the warm air, which in turn is lifted and moves northeastward aloft into the region occupied by the colder air. The leading edge of the northeastward-moving warm air is a warm front. The leading edge of the southeastward-moving cold air behind the warm air is a cold front. The rising warm air is usually associated with a

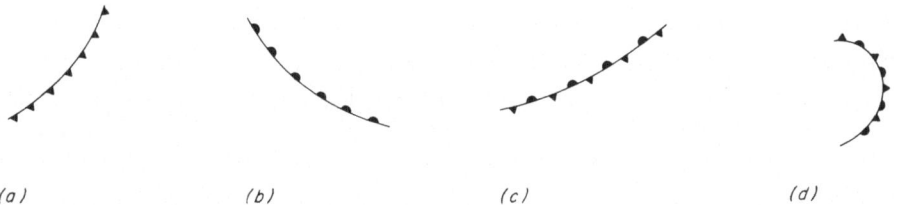

Fig. VII-29. *Fronts.* Conventional symbols used to denote (*a*) cold front, (*b*) warm front, (*c*) stationary front, and (*d*) occluded front.

leading surface low pressure centre and the sinking cold air behind is usually associated with a following surface high pressure centre. As the system develops, the low pressure centre deepens. The surface isobars often become closed, as shown in Figure 30(c).

As the cold front cuts in under the warm air, and the warm air is lifted and escapes to higher levels over the warm front, the warm sector narrows. The cold front tends to overtake the warm front; this process is called *occlusion*. The front resulting from the combination of the warm and the cold front is called an *occluded front* (Figure 30(d)).

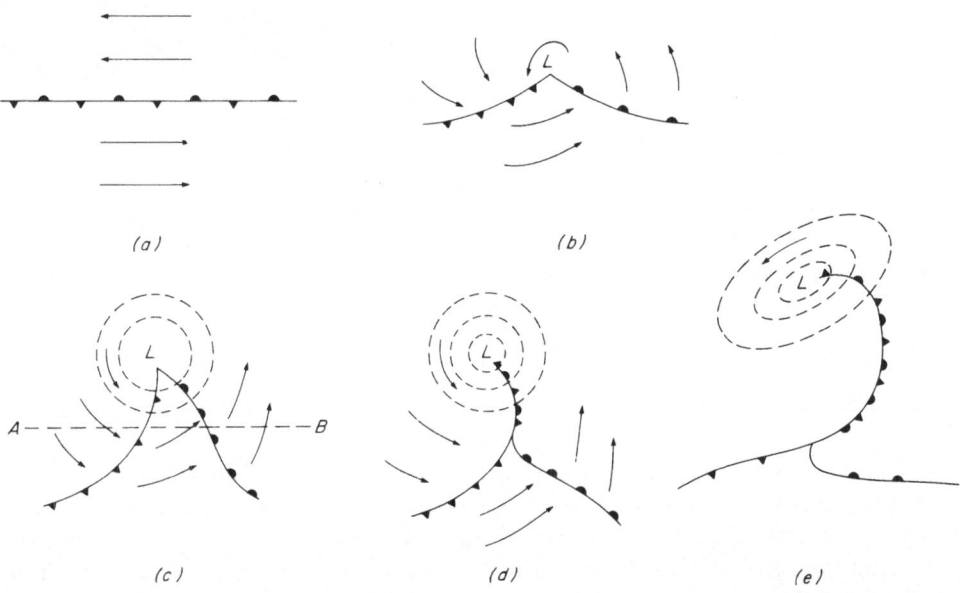

Fig. VII-30. *Life cycle of extratropical cyclone in the Northern Hemisphere.* (*a*) Midlatitude cyclones usually develop on the more or less stationary polar front. (*b*) As the warm air south of the polar front is pushed northward, it tends to rise and to move northeastward while the cold air cuts southeastward under the warm air from behind. The leading edge of the northeastward-moving warm air forms a warm front, while the leading edge of the cold air moving southeastward forms a cold front. (*c*) The surface pressure at the tip of the frontal system drops and closed isobars are often formed. (*d*) As the system develops, part of the warm air is lifted off the ground; the occlusion process starts. The boundary region between the cold air masses in front of and behind the system is called an occlusion front. (*e*) When most of the intruding warm air is lifted to the upper levels, the system has exhausted its source of energy (available potential energy). It reaches a dissolving stage. The system will be dissipated soon afterwards.

189

It is the boundary between the cold air ahead of the warm front and the cold air behind the cold front. When most of the warm air has been lifted to the upper levels, the occlusion process is completed, as shown in Figure 30(e). At this stage, all the potential energy that is available for conversion into kinetic energy has been exhausted. The disturbance reaches the end of its development. The kinetic energy of the system will be quickly dissipated soon afterwards.

During the life cycle of the disturbance, warm air moves poleward and cold air moves southward (Northern Hemisphere). A northward transport of heat energy is accomplished.

To illustrate the vertical structure of a cyclone, the vertical cross section cut along the line *AB* in Figure 30(c) is shown schematically in Figure 31. Both the warm front and the cold front are sloped with the warm air lying on top of the cold air. The cold front is usually steeper than the warm front. The figure also shows that the warm air glides over the cold air along the warm front, and the cold air cuts under the warm air behind the cold front.

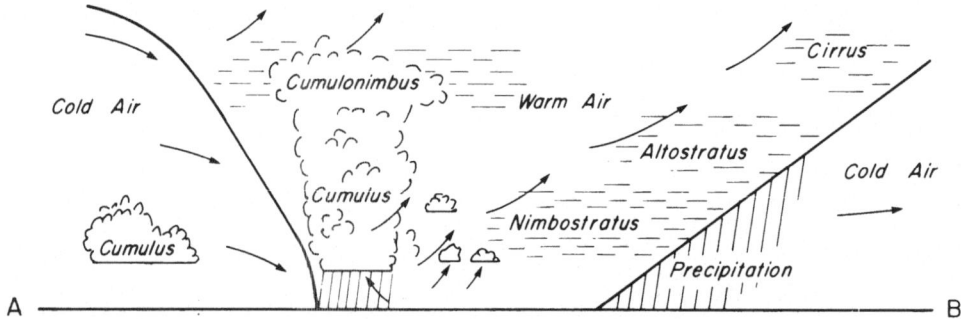

Fig. VII-31. *Vertical cross section of a frontal system.* The vertical cross section of a mid-latitude cyclone along line AB in Figure 30(c) is shown. The cold front is usually steeper than the warm front. The sequence of cirrus, altostratus and nimbostratus clouds are the typical signs of an approaching warm front. Due to the strong uplifting by the cold air, convective clouds often develop along the cold front. Behind the cold front, the sky is relatively clear, because of the general downward air motion.

The warm air is usually of a tropical or subtropical origin. It often has a high moisture content. As it is lifted and moving northeastward over the warm front, condensation often occurs and various types of clouds will form. The first sign of an approaching warm front is the appearance of cirrus clouds at a very high level. As the front continues to approach, the clouds become lower and thicker. Altostratus clouds associated with drizzle will appear, and then give way to nimbostratus with heavier rain as the front gets nearer. This distribution of clouds over a warm front is schematically illustrated in Figure 31.

In the warm sector of the disturbance, the amount of cloud and precipitation depends very much on the characteristics of the warm air mass. Sometimes, stratus with drizzle may occur; sometimes cumulus clouds may develop, depending on the vertical stability of the air mass.

There is usually a drastic change in weather conditions as the cold front approaches. Because of the strong uplifting from below by the cold air, vertical instability is more likely to develop. Cumulonimbus clouds often appear, with heavy precipitation (see Figure 31). As the cold front passes by, air temperature decreases drastically. Weather conditions behind the cold front depend again on the characteristics of the cold air mass.

Because of the general downward motion of air, the sky is relatively clear. Convection in the form of cumulus cloud sometimes develops, producing showers.

The occlusion process has also characteristic patterns of clouds and precipitation. Both stratus and cumulus clouds may appear, producing either steady or intermittent showers.

The development of mid-latitude cyclones and the positions of fronts can often be identified in satellite pictures. One such example is given in Figure 32 which shows the picture in the visible range of a cyclone located at the east coast of Newfoundland and Nova Scotia, Canada, at 1247 Z,* 10 September 1978. The long, narrow, curved cloud band marks the position of the cold front. The cyclone at the time shown had already been at the occlusion stage for about 24 hours. The low pressure centre was located at about 50°W and 45°N. More or less regularly arranged, smaller-scale cloud patterns behind (that is, to the west of) the cold front are associated with convection caused by vertical instability. As a cold air mass moves over a warm ocean surface, the air often becomes vertically unstable; convection usually develops, forming clouds which are often regularly arranged.

The small, bright cloud mass at the lower right-hand corner of the picture is Hurricane Flossie at its initiation stage.

The surface pressure distribution and the positions of the fronts of the system at 1200 Z, 10 September, 1978 is given in Figure 33. The surface pressure at the low pressure centre was about 988 mb. The tip of the warm and cold front system was already separated from the low pressure centre, indicating that the system was very much occluded. No occluded front, however, is shown in the map because no such frontal contrast is noticeable at the surface.

As the occlusion process progressed, the low pressure centre deepened further. The satellite picture of the system taken at 1206 Z, 11 September 1978 and the accompanying surface map at 1200 Z, 11 September 1978 are given in Figures 34 and 35, respectively. The low pressure centre had moved northeastward to the position of 45°W and 52°N; pressure at the centre had now dropped to 985 mb. The system had more or less reached the end of its life cycle and begun to dissolve. Again, no occluded front is shown in the surface map. The thick, barbed, curved line starting from the tip of the cold and warm frontal system toward the low pressure centre is called a *trowal*, meaning a *trough of warm air aloft*. At this time, a pocket of occluded warm air was located above the cold air over the low pressure area.

It is interesting to note from Figure 34 that Hurricane Flossie had now moved into the frontal region marked by the narrow cloud band. This usually has the effect of further intensifying the development of the hurricane, because additional potential energy across the front is available for transformation into kinetic energy of air motion.

---

* That is, 12 hours and 47 minutes Greenwich Mean Time.

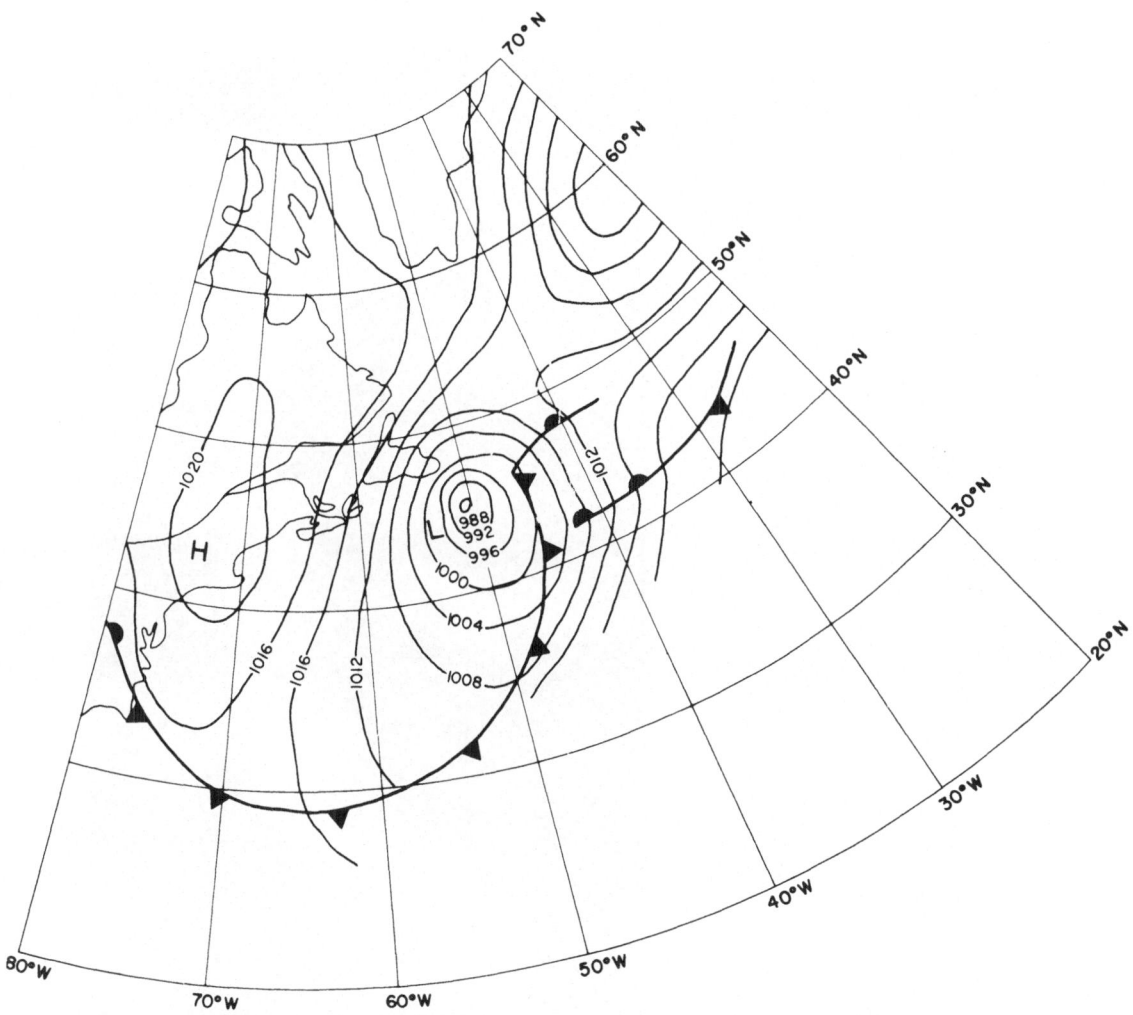

Fig. VII-33. *Surface map*, at 1200 Z, 10 September 1978, of pressure distributions and positions of fronts of the cyclone shown in Figure 32.

Fig. VII-32. *Middle-latitude cyclone*. NOAA5 satellite picture of a mid-latitude cyclone taken in the visible range of the radiation spectrum at 1247 Z, 10 September 1978. The *X* signs on the picture mark the positions of latitudinal and longitudinal circles at 10° intervals. The centre of the low pressure system was located at about 50°W and 45°N. The cyclone was in an occluded stage.

193

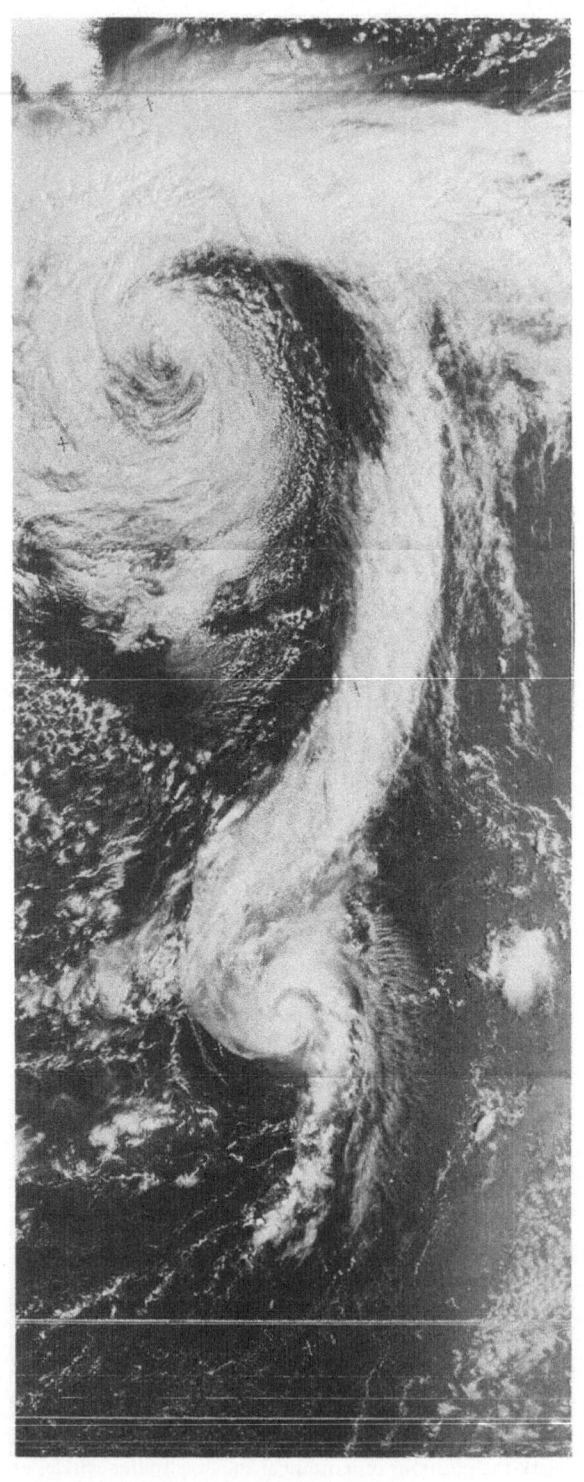

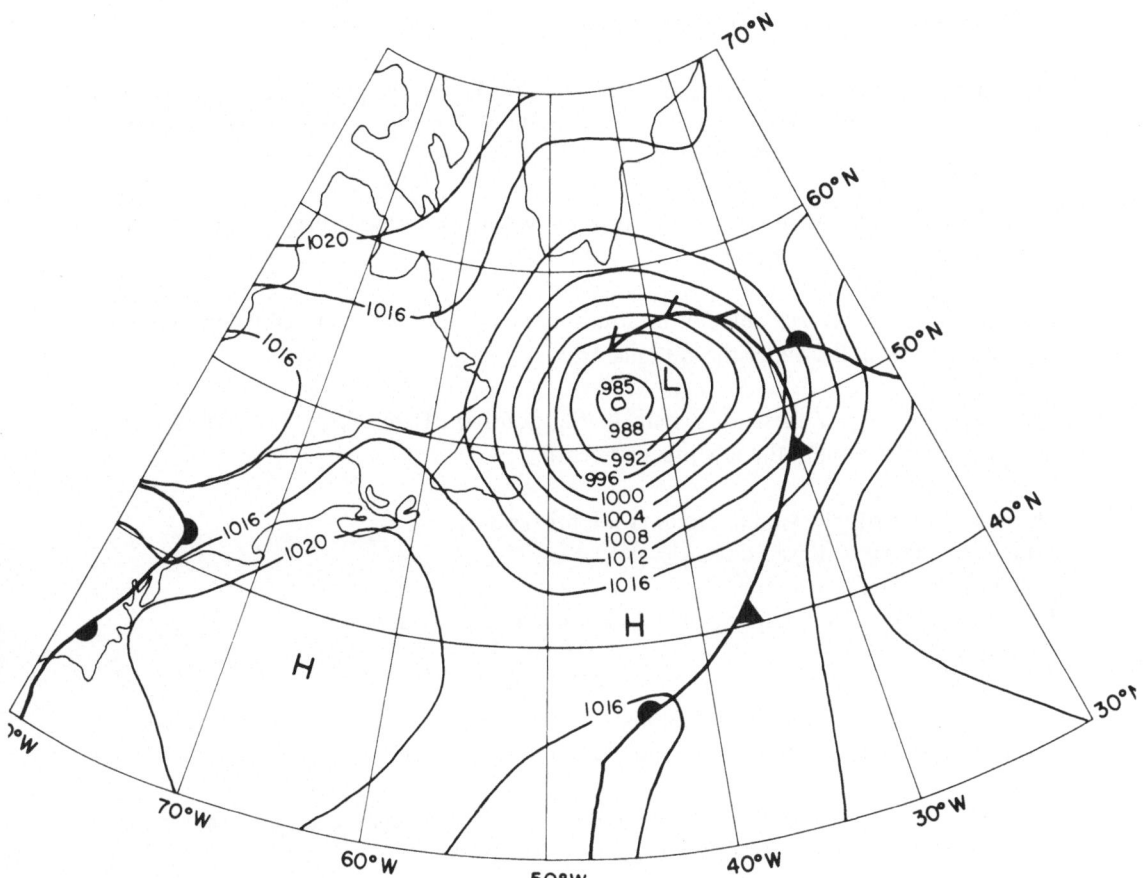

Fig. VII-35. Same as Figure 33, except at 1200 Z, 11 September 1978.

Fig. VII-34. Same as Figure 32, except taken at 1206 Z, 11 September 1978. The low pressure centre was now located at 52°W and 45°N. The cyclone had reached a dissolving stage.

Chapter VII: Questions

Q1. Consider a cylindrical air parcel. The velocity field in the parcel is the superposition of an isotropic contraction and a rigid body rotation. The angular velocity of the cylinder should increase with time. Why?

Q2. The thickness of an atmospheric layer between two isobaric surfaces is proportional to its mean temperature. Why?

Q3. The typical vertical velocity in atmospheric motion systems usually decreases with increasing horizontal scale of the systems. Why?

Q4. Consider a low pressure system in the Southern Hemisphere. In which direction does air circulate around the low pressure centre?

Q5. The geostrophic wind approximation breaks down
 (*i*) in equatorial regions
 (*ii*) near the Earth surface
(*iii*) if the isobaric lines are curved and have small radii of curvature.
Explain why.

Q6. Near the Earth's surface, the observed wind often has a component directed from the high pressure region toward the low pressure region. Why?

Q7. The strength of the westerly wind in middle latitude regions is a good indication of the strength of the north–south temperature gradient. Why?

Q8. When we have warm advection, i.e. air blows from a warm region toward a cold region, the geostrophic wind rotates clockwise with height. Why?

Q9. When potential energy is being converted into kinetic energy in a large-scale weather system, the internal thermal energy of the system also decreases. Why?

Q10. During the winter season, continental air masses tend to be much colder than maritime air masses. Why?

Q11. Mid-latitude cyclones always produce precipitation. Why?

Q12. Mid-latitude cyclones usually develop along the polar front. Why?

Chapter VII: Problems

(Any necessary constants not given in the statement of a problem will be found in the Table of Constants on pages x–xi)

P1. The wind field in a region of the atmosphere is given by

$$\mathbf{V} = \hat{x}(a_0 + a_1 x + a_2 y) + \hat{y}(b_0 + b_1 x + a_1 y)$$

Determine the acceleration of the air parcel located at $x = 0$ and $y = 0$.

P2. Show that the wind field given in **P1**, above, is composed of a linear translation, an isotropic expansion and two simple shear motions.

P3. An air parcel on the Earth's surface moves northward horizontally from the equator without interacting with the rest of the atmosphere. Let us assume that the initial east–west velocity of the parcel relative to the Earth is zero. Determine its east–west velocity relative to the Earth when it reaches 45°N.

P4. The atmosphere is said to be incompressible if $D\rho/Dt = 0$. Show that in this case the vertical velocity can be determined from the horizontal wind distribution.

P5. 100 Joules of heat energy is added to a hydrostatic air column above the Earth's surface. Determine the increase in internal energy of the air column.

P6. The surface pressure in a region of the atmosphere is 1000 mb. Suppose that the 900 mb surface is at a height of 1.2 km. Estimate the mean temperature of the air between the 1000 and 900 mb surfaces.

P7. Calculate the geostrophic wind speed for a pressure gradient of 0.03 mb/km. Let $\rho = 10^{-3}\,\text{gm cm}^{-3}$ and $f = 10^{-4}\,\text{sec}^{-1}$.

P8. Eddy motions in the atmosphere may transport horizontal momentum in the north–south direction. Construct the streamlines of an eddy motion that is capable of doing this.

Bibliography

The following list is not meant as an exhaustive bibliography, but as an indication for readers who wish to learn more about some particular field, or simply to complement a course with parallel reading. It has been divided into several sections: general books and specialized monographs. In the first part, a number of books have been included which are purely or predominantly meteorological; this has been indicated in the comments. The references are ordered within each section according to the level, starting with the most elementary texts.

General

R. A. Anthes, H. A. Panofsky, J. J. Cair and A. Rango: *The Atmosphere*. Merrill, 1975. 339 pp.
  Purely meteorological in approach. Descriptive, on very elementary level, as for any kind of reader. It includes chapters on atmospheric optics, atmospheric pollution and climate change.
H. Flohn: *Weather and Climate*. McGraw-Hill, 1969. 253 pp.
  Elementary, with a purely meteorological approach.
R. H. Goody and J. C. G. Walker: *Atmospheres*. Prentice-Hall, 1972. 150 pp.
  Excellent text with a modern general approach on the atmosphere. Elementary and mostly descriptive. It includes a chapter on the evolution of atmospheres. It provides ideal parallel reading for a course.
G. M. B. Dobson: *Exploring the Atmosphere*. Oxford, at the Clarendon Press, 2nd ed., 1968. 209 pp.
  Emphasis on subjects of high atmosphere (7 chapters out of 11).
L. J. Battan: *Fundamentals of Meteorology*. Prentice-Hall, 1979. 336 pp.
J. G. Harvey: *Atmosphere and Ocean: our Fluid Environments*. The Artemis Press, Sussex, 1976. 143 pp.
  Mostly meteorological and includes oceanic circulations. Elementary.
R. G. Barry and R. J. Chorley: *Atmosphere, Weather and Climate*. Methuen, 3rd ed., 1976. 419 pp.
  Mainly meteorological and climatological, i.e. stressing all aspects related to weather and climate. Aimed at geography students.
J. M. Wallace and P. V. Hobbs: *Atmospheric Science*. Academic Press, 1977. 467 pp.
  This textbook covers much of the material included in the present one, but is on a higher level and more extensive, as directed to students specializing in the field. It has a large collection of problems.
S. L. Hess: *Introduction to Theoretical Meteorology*. Holt, Rinehart and Winston, 1959. 362 pp.
R. G. Fleagle and J. A. Businger: *Introduction to Atmospheric Physics*. Academic Press, 1963. 346 pp.
J. T. Houghton: *The Physics of Atmospheres*. Cambridge University Press, 1977. 203 pp.
  Advanced textbook on general meteorology, suitable for senior undergraduate or graduate students specializing in the field.

Atmospheric Chemistry

S. S. Butcher and R. J. Charlson: *An Introduction to Air Chemistry*. Academic Press, 1972. 241 pp.
B. H. Svensson and R. Söderlund (eds.): Nitrogen, Phosphorus and Sulfur – Global Cycles. Ecological Bulletins No. 22, Royal Swedish Academy of Sciences 1975. R. Söderlund and B. H. Svensson: *The Global Nitrogen Cycle*. L. Granat, R. O. Hallberg and H. Rodhe: *The Global Sulfur Cycle*.
C. E. Junge: *Air Chemistry and Radioactivity*. Academic Press, 1963. 382 pp.
  Classical monograph on the subject.

Radiation and Upper Atmosphere

R. A. Craig: *The Upper Atmosphere*. Academic Press, 1965. 509 pp.

C. O. Hines, I. Paghis, T. R. Hartz and J. A. Fejer (eds.): *Physics of the Earth's Upper Atmosphere*. Prentice-Hall, 1965. 434 pp.

K. L. Coulson: *Solar and Terrestrial Radiation*. Academic Press, 1975. 322 pp.

K. Ya. Kondrat'yev: *Radiative Heat Exchange in the Atmosphere*. Pergamon Press, 1965. 411 pp.

R. Goody: *Atmospheric Radiation. I-Theoretical Basis*. Oxford, at the Clarendon Press, 1964. 436 pp.

Thermodynamics

N. R. Beers: 'Meteorological Thermodynamics and Atmospheric Statics'. In F. A. Berry, E. Bollay and N. R. Beers: *Handbook of Meteorology*. McGraw-Hill, 1945, pp. 313–409.

J. V. Iribarne and W. L. Godson: *Atmospheric Thermodynamics*. D. Reidel, 1973. 222 pp.

See also corresponding chapters in textbooks on Dynamical Meteorology.

Cloud Physics

H. R. Byers: *Elements of Cloud Physics*. University of Chicago Press, 1965. 191 pp.

R. R. Rogers: *A Short Course in Cloud Physics*. Pergamon Press, 2nd ed., 1978.

B. J. Mason: *Clouds, Rain and Rainmaking*. Cambridge University Press, 1962. 145 pp.

N. H. Fletcher: *The Physics of Rainclouds*. Cambridge University Press, 1962. 386 pp.

B. J. Mason: *The Physics of Clouds*. Oxford, at the Clarendon Press, 2nd ed., 1971. 671 pp.
Classical reference monograph on the subject.

H. R. Pruppacher and J. D. Klett: *Microphysics of Clouds and Precipitation*. D. Reidel, 1978. 706 pp.
Extensive up-to-date monograph, covering most microphysical aspects and giving condensed summaries of the basic physics involved.

Atmospheric Electricity

J. A. Chalmers: *Atmospheric Electricity*. Pergamon Press, 2nd ed., 1967. 515 pp.

H. Israel: *Atmospheric Electricity*. Israel Program for Scientific Translations, 1971. 2 vols. 796 pp.

M. A. Uman: *Lightning*. McGraw-Hill, 1969. 264 pp.

Atmospheric Dynamics and Meteorology

J. R. Holton: *An Introduction to Dynamic Meteorology*. Academic Press, 1972. 319 pp.
Intermediate level. Atmospheric dynamics is developed in a logical fashion and treated as a coherent subject.

S. Petterssen: *Weather Analysis and Forecasting*. McGraw-Hill, 1956. Vol. I, 446 pp. and Vol. II, 266 pp.

G. J. Haltiner and F. L. Martin: *Dynamical and Physical Meteorology*. McGraw-Hill, 1957. 470 pp.

E. Palmén and C. W. Newton: *Atmospheric Circulation Systems*. Academic Press, 1969. 603 pp.
Advanced. A very comprehensive monograph on atmospheric circulation systems.

E. N. Lorenz: *The Nature and Theory of the General Circulation of the Atmosphere*. World Meteorological Organization, 1967.

Answers to Selected Questions and to the Problems and Indications for their Resolution

Chapter I

Q2. Below 100 km (homosphere), air is stirred by turbulent motions and overturnings. Above 100 km (heterosphere) there are no such motions, and the vapour will diffuse gradually.

Q4. Because of the high rate of ionization by the solar radiation, at those levels. (*i*) At higher altitudes molecules become more scarce. (*ii*) At lower levels the appropriate ionizing radiation has been exhausted.

P1. 5.27×10^{18} kg

If m = mass of earth, p = pressure, ρ = density, z = height,

$$p = \int_0^\infty g\rho \, dz = g \int_0^\infty \rho \, dz = g \cdot m$$

P2. 863 m

Integrate hydrostatic equation.

P3. (*a*) 1536 kg/m²; (*b*) 1.19 km.

(*a*) apply hydrostatic equation; (*b*) apply gas law.

P4. (*a*) $g = g_0(1 - 3.14 \times 10^{-7} z)$ (z in m); (*b*) $m = 5.96 \times 10^{24}$ kg.

Apply law of gravitational attraction.

P5. 1638 km.

Use partial pressure distributions, with proper scale heights.

P6. 38 km.

As **P5**.

P7. (*a*) 2.5×10^{19} cm^{-3}; (*b*) 3.2×10^{13} cm^{-3}; (*c*) 1.4×10^8 cm^{-3}.

Apply gas law.

P8. The electronic density is 1.0×10^6 cm^{-3} at 300 km of altitude.

P9. At low frequencies only the emitted signal will appear: the wave pulses are not reflected. When the frequency increases, a peak (reflected signal) will appear to the right, at a distance increasing with the height of the reflection. As the frequency continues increasing, the second peak is gradually displaced toward the right (electron densities increasing with increasing height).

Continuing the increase in frequency, the echo disappears and a second echo, more to the right of the screen, is observed: the maximum of E was reached, and now the pulses are reflected higher up, in the F region. With further increase of frequency, this second echo also becomes displaced to the right (electron density increasing with height, in the F region), until it again disappears (at the maximum of F).

P10. (*a*) As the particle moves in the field, it is subjected to the electromagnetic deflecting force, and consequently to the centrifugal force. At equilibrium, these two forces balance and the particle describes a circumference in a plane perpendicular to **B**. Its radius may be obtained from the balance condition:

$$evB = \frac{mv^2}{r}$$

$$r = \frac{mv}{eB} = \text{const}$$

where m is the mass and e the charge of the particle.

(*b*) The electromagnetic or Lorenz force **F** only acts on the common component perpendicular to **B**. The result is a spiral of radius $r = mv_\perp/eB$ with a separation between turns of $2\pi m v_\parallel/eB$ (the rotation period being $2\pi m/eB$). $v_\perp$ and $v_\parallel$ are here the components of **v** perpendicular and parallel to **B**.

Chapter II

Q4. CO_2 and O_2.

Q7. SO_2 will decrease the pH because of the equilibria:

$$SO_2 + H_2O \rightleftarrows H_2SO_3 \overset{\rightarrow}{\leftarrow} H^+ + HSO_3^-$$

NH_3 will increase the pH because of the equilibria:

$$NH_3 + H_2O \rightleftarrows NH_4OH \rightleftarrows NH_4^+ + OH^-$$

$$OH^- + H^+ \rightleftarrows H_2O$$

Q9. Yes, because the H^+ produced in the equilibria

$$CO_2 + H_2O \rightleftarrows H_2CO_3 \rightleftarrows H^+ + HCO_3^-$$

and the OH^- produced by the dissolved ammonia:

$$NH_3 + H_2O \rightleftarrows NH_4OH \rightleftarrows NH_4^+ + OH^-$$

will combine into H_2O. The decrease in concentration of OH^- will displace these equilibria towards the right (law of mass action).

P1. 6.6×10^{13} kg.

Molar ratio = volume ratio. Therefore

$$18\,\text{ppm} = \frac{18 \times 10^{-6}}{1} = \text{molar ratio}$$

$$\text{Mass ratio} = \frac{18 \times 10^{-6} \times A_{Ne}}{1 \times \overline{M}_{air}}$$

where A_{Ne} = atomic weight of Ne and $\overline{M}_{air}$ = average molecular weight of air. This ratio, multiplied by the mass of the atmosphere, will give the answer.

P2.
- N_2 75.52% in mass
- O_2 23.15% in mass
- Ar   1.28% in mass
- CO_2 0.05% in mass

Cf. **P1**.

P3. 1 ppm = $2857\,\mu g/m^3$; $1\,\mu g/m^3 = 3.50 \times 10^{-4}$ ppm
Cf. **P1**. 1 mol of ideal gas at STP occupies $0.0224\,m^3$.

P4. (a) 2.04 mm; (b) 0.204 mm; (c) 4×10^{-3} mm.
Cf. **P I-1**.

P5. (a) 1.76 g/kg; (b) 2830 ppm.
The volume mixing ratio is equal to the ratio of partial pressures: $p_s/(p - p_s)$ (p_s = saturation vapour pressure; p = total pressure).

P6. (a) 20.34 g; (b) 0.12 g; (c) 3.8×10^{-3} g.
Cf. **P1** and **P5**.

P7. 6.0 mb.
Apply Equation (2).

P8. 15 days.
Cf. **P1**. Apply Equation (1).

P9. 3.88 kg.

P10. 4 to 8 days.
Cf. **P8**.

P11. $10\,cm^{-3}$.
Integrate Equation (10).

Chapter III

Q10. The long wave radiation lost to space would be that of a black body at the temperature of the highest thin layer, at about the tropopause level. This would be less than the total radiation lost by the Earth in the real case (cf. Figure 13). Therefore the temperature of the tropopause should increase until it again balances the incoming solar radiation, i.e. until it becomes equal to the effective temperature of the Earth.

P1. (a) $1.30 \times 10^{15} \, s^{-1}$; (b) $8.63 \times 10^{-19} \, J$; (c) $520 \, kJ/mol$;
(d) Yes.

P2. 20.2 W.

P3. (a) $9.8 \times 10^{13} \, J/s.m^3 = 9.8 \, kW/cm^2 . \mu m$
(b) $2.9 \, kW/cm^2. \mu m$.

P4. Write Planck's law in the form

$$E_\lambda = \lambda^{-5} f(\lambda T)$$

Find the maximum condition by differentiating with respect to λ and equating to 0. The resulting condition can be written

$$F(\lambda T) = 0$$

which is satisfied for a value of the product λT.

P5. $10 \, \mu m$.
Apply Wien's displacement law (7).

P6. 5900 K.
Apply Equation (5) and surface area of Sun, and consider solid angle.

P7. (a) $-55°C$; (b) $-90°C$.
Radiative equilibrium: As much is absorbed per unit time (incident radiation times absorptivity) as is emitted (Equation 5).

P8. (a) $T_1 = (E/\sigma)^{1/4}$; $T_i = i^{1/4} T_1$; $T_0 = (n+1)^{1/4} T_1$;
(b) $T_0 = 440 \, K = 167°C$.
Consider radiation balance above top plate, and then radiative equilibrium successively for each plate, starting at the top:

Balance above Plate 1: $E = \sigma T_1^4$

Plate 1: $2\sigma T_1^4 = \sigma T_2^4 \quad T_2 = 2^{1/4} T_1$

Plate 2: $2\sigma T_2^4 = \sigma T_1^4 + \sigma T_3^4$, where T_2 can be replaced by previous expression, and then solved for T_3.

etc.

P9. (a) 0.5; (b) 0.575.
(a) Consider incidence angles; (b) Equation (5).

P10. 1.65%.
In the radiative equilibrium, the emitted power balances the power received, and this is inversely proportional to the square of the distance Sun–Earth. Differentiate logarithmically the corresponding equation.

Chapter IV

Q4. The water vapour partial pressure is proportional to the total pressure, according to

$$e = N_v p$$

where N_v is the molar fraction of water vapour.

Q12. Soon after the parcel reaches the level at which its temperature has become equal to that of the environment.

Q14. In the first case, the atmosphere has a stable temperature stratification, whereas in the second case the air may be close to neutral stratification ($\beta \cong \beta_d$), so that the temperature of the effluent gases is enough to produce turbulent convection.

P1. 0.0234.

From the figure, $e_s = 23.4$ mb.

N_v = ratio of partial pressure to total pressure.

P2. $T_d = 258.8$ K $= -14.4°$C; $T_f = 260.2$ K $= -12.9°$C.

Starting from the saturation vapour pressure at $0°$C (6.1 mb, common to water and ice), integrate the Clausius–Clapeyron equation to 2 mb along the ice-vapour and the ice-water equilibrium curves.

P3. 27.9 J/kg.min

Use Equation (20).

P4. (a) 0.042 K/min; (b) 7.0°C; (c) 0.011 K/min.

(a) Use Equation (18). (c) Use Equation (20).

P5. 5.9 g/m³.

Use gas law.

P6. 21 kJ/kg.

See Equation (12), cf. **P1**.

P7. 46.4%.

Consider Equation (23) and that because $e = N_v p$ is proportional to p, a similar equation can be written substituting e for p. Apply this equation between initial and final states.

P8. (a) $-33°$C; (b) 20°C; (c) 5.6 km; (d) $-13.7°$C.

(a) Apply Equation (23). (b) See Equation (24). (c) Integrate the hydrostatic equation:

$$\ln\left(p_0/p\right) = \int_0^z \frac{gM}{RT}\, dz$$

In the integral, $T = T_0 - \beta z$. (d) Use the previous expression for T.

P9. (a) -3.15×10^{-4}; (b) 0.37 g/m³.

(a) Apply Equation (27). (b) Cf. **P1** and **P5**.

P10. (a) 87% and 55%, respectively. (b) Yes, a fog is formed.

No; condensation releases latent heat of vaporization, which modifies the final temperature and the corresponding saturation vapour pressure.

P11. It can be imagined that the air of the whole layer is brought adiabatically to 1000 mb, in which each infinitesimal layer acquires the corresponding potential temperature θ (by definition). The mixture, at 1000 mb, will give a uniform temperature which is the weighted average. The mass of an infinitesimal layer is (per unit cross section)

$$\rho\, dz = -\frac{1}{g}\, dp$$

and the average will be

$$\frac{\int_{z_1}^{z_2} \theta\rho\, dz}{\int_{z_1}^{z_2} \rho\, dz} = \frac{\int_{p_2}^{p_1} \theta\, dp}{p_1 - p_2}$$

This is $\bar{\theta}$, because it is the temperature of the whole mass at 1000 mb. The mass may now be redistributed between p_1 and p_2 adiabatically, thus conserving the value $\bar{\theta}$ throughout the resulting layer.

P12. (i) (a) 945 mb; (b) 30°C and 28.8°C; (c) 12 K/km; unstable; convection ensues.

(ii) (a) 943 mb; (b) 20°C and 21.9°C; (c) 6 K/km; stable.

(d) $d\theta/dz < 0$: unstable; $d\theta/dz > 0$: stable.

(a) Integrate hydrostatic equation. (b) Apply Equation (24); $\theta_0 = T_0$.

(c) Compare β with $\beta_d = 9.8$ K/km. (d) Consider that θ is constant through height for $\beta = \beta_d$, and the definition of $\theta = T\,(1000 \text{ mb}/p)^{R/C_p}$.

Chapter V

Q3. It is unstable. If a slight evaporation occurs, the vapour pressure of the drop increases, and therefore tends to further evaporate. The reverse is true for a slight condensation.

Q4. Freezing point depression, boiling point increase, osmotic pressure (see textbooks on Physical Chemistry).

Q8. Higher velocities will lead to higher supersaturations. This depends on the competition between rate of cooling and rate of condensation. When the air rises from one level to another close above, the increase in supersaturation is controlled in principle by the decrease in temperature; however, it is partially compensated by the condensation. If the velocity is higher, the time for condensation is shorter and the supersaturation has increased more.

The rate of condensation depends also at each level on the concentration of activated CCN. A level will finally be reached at which this concentration prevents further increases in supersaturation; the maximum supersaturation attained will be higher for higher updraught speeds, due to the previous effect acting at all levels.

Q9. Considering Figure 11, it may be readily seen that the terminal velocity V is proportional to r^2 (r = radius) in the first part of the curve, and tends to become constant towards 3 mm radius.

Q13. The idea is to stimulate the collision-coalescence process by producing large drops. The nuclei should be large enough to produce drops which fall appreciably faster than most cloud droplets and thus grow within the updraught, but not so large as to fall rapidly out of the cloud.

P1. (a) $1.9\,\mu$m; (b) $3.3\,\mu$m.

(a) Combine Equations (1) and (2). (b) Find maximum of the vapour pressure as a function of r, by differentiating and equating to 0.

P2. (a) 31.7 mb; (b) 0.19%.

(a) Cf. **P1**.

P3. (b) 52s; 187s; 750s

(a) Plot Equation (4). (b) Use $t = (R_2^2 - R_1^2)/2C$, from integrating (4).

P4. $v = \dfrac{2g\rho}{9\eta} r^2 = 1.3 \times 10^{-2} r^2$ (r in μm, v in cm/s).

Equate weight and drag force.

P5. (b) 665s.

Use Equation (7).

P6. (b) 100 min.

Use Equations (4) and (7).

P7. (a) $R = 1.0$ mm; (b) $\Delta z = 0.50$ km.

(a) Use Equation (8) for F. Integrate Equation (13) and solve for R by successive approximations. (b) Differentiate (12) and equate to 0 to find R for the top of the trajectory. Introduce this value into (12) to find Δz.

Chapter VI

Q3. Being moving charged particles, they will be deflected by the magnetic field of the Earth. This will result in a latitude dependence.

Q5. The ionization potential of the neutral molecules in the first case, and their electron affinities in the second case.

Q6. Because of the attraction between the charge of the ion and the dipole moments (induced or permanent) of the neutral molecules.

Q8. In polluted air, Equation (10) becomes

$$q \simeq \beta n \mathrm{N}$$

so that $n \propto \mathrm{N}^{-1}$.

Q11. Electrification mechanisms such as described in §6 cannot develop enough in shallow layer clouds.

P1. When the plates are exposed to the atmospheric field, a surface charge of density $\sigma = \epsilon_0 E$ is induced, that returns to ground when the plates are covered. This charging and discharging is repeated with the frequency of covering and uncovering.

P2. (a) $E = -38.44 - 81.56 e^{-0.00452 \cdot z}$ (V/m); $V = 38.44 \cdot z + 1.804 \times 10^4 (1 - e^{-0.00452 \cdot z})$ (V).
(b) $8.3\,\text{cm}^{-3}$; $-71.5\,\text{V/m}$; $18\,423\,\text{V}$.
(a) Integrate Equation (4) and apply Equation (9).

P3. $16\,\text{pC/m}^3$.

P4. $20\,\text{pC/m}^3$.

Apply $V = -\int_R^0 E\,dr$, where R is the radius of the cage.

P5. (a) $877\,\text{cm}^{-3}$; (b) 0.123 and 0.877, respectively;
(c) $2469\,\text{cm}^{-3}$; 0.975 and 0.025, respectively.
Apply Equation (10).

P6. $4.8 \times 10^{-14}\,\Omega^{-1} m^{-1}$
$j = \lambda E = n_+ e v_+ + n_- e v_- = e(n_+ + n_-) kE$

P7. (a) $Q = Q_0 \exp\left(-\dfrac{\lambda}{2\epsilon_0} t\right)$, where λ is the conductivity of the air.

(b) 15 min.

(a) $-dQ/dt = 4\pi R^2 \dfrac{\lambda}{2} E$ (only ions of opposite sign to Q contribute to the discharge).

P8. (a) $72\,\text{kV/m}$; (b) $900\,\text{kV/m}$.

P9. (a) $10^6\,\text{MW}$; (b) $10^4\,\text{MW}$.
Power $= VQ/t$.

Chapter VII

Q1. Because angular momentum must be conserved.

Q3. $w/u \leqslant D/L$.
where w = typical vertical velocity, u = typical horizontal velocity, D = vertical length scale, L = horizontal length scale.
Usually u does not vary greatly with L, and D is bounded by the depth of the troposphere. Therefore w usually decreases with increasing L.

Q5. (i) The Coriolis parameter, and hence the Coriolis force, becomes very small when one approaches the equatorial region.
(ii) The surface frictional force becomes important.
(iii) Centripetal acceleration is no longer negligible.

Q6. Due to the presence of surface frictional force.

Q7. Thermal wind relation.

Q8. Due to thermal wind. See Figure 15.

P1. Using $\dfrac{D\mathbf{V}}{Dt} = \dfrac{\partial \mathbf{V}}{\partial t} + \mathbf{V}\cdot\nabla\mathbf{V}$, at $x = 0, y = 0$

$$\frac{D\mathbf{V}}{Dt} = \hat{x}(a_1 a_0 + a_2 b_0) + \hat{y}(b_1 a_0 + a_1 b_0)$$

P2. Linear translation: $\mathbf{V}_1 = \hat{x}a_0 + \hat{y}b_0$
Isotropic expansion: $\mathbf{V}_2 = \hat{x}a_1 x + \hat{y}a_1 y$
Simple shear motions: $\mathbf{V}_3 = \hat{x}a_2 y$
$$\mathbf{V}_4 = \hat{y}b_1 x$$

P3. Use conservation of angular momentum

$$(\Omega R \cos\phi + u)R\cos\phi = \text{constant}$$

where ϕ = latitude, R = radius of Earth, Ω = Earth's angular velocity, u = zonal wind speed. At 45°,

$$u = \frac{\Omega R}{\cos 45°} - \Omega R \cos 45° = 327\,\text{m/s}$$

P4. If $D\rho/Dt = 0$, the continuity equation becomes

$$\nabla \cdot \mathbf{V} = 0$$

$$\text{or } \frac{\partial u}{\partial x} + \frac{\partial v}{\partial y} + \frac{\partial w}{\partial z} = 0$$

$$w = -\int_0^z \left(\frac{\partial u}{\partial x} + \frac{\partial v}{\partial y}\right) dz$$

P5. 71.4 Joule.

P6. 389.3 K.

P7. 30 m/s.

P8.

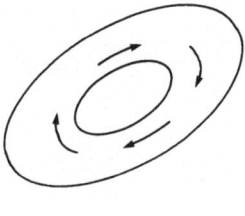

(a)   **(b)**

Index